D0092905

LIVE SUSTAINABLY NOW

KARL COPLAN

LIVE SUSTAINABLY NOW

A Low-Carbon Vision of the Good Life

COLUMBIA UNIVERSITY PRESS
NEW YORK

Columbia University Press
Publishers Since 1893
New York   Chichester, West Sussex
cup.columbia.edu

Library of Congress Cataloging-in-Publication Data
Names: Coplan, Karl S., author.
Title: Live sustainably now : a low-carbon vision of the good life / Karl Coplan.
Description: New York : Columbia University Press, [2020] | Includes bibliographical references and index.
Identifiers: LCCN 2019019478 | ISBN 9780231190909 (cloth : alk. paper) | ISBN 9780231549165 (ebook)
Subjects: LCSH: Sustainable living—United States. | Energy conservation—United States. | Water conservation—United States. | Environmental responsibility—United States.
Classification: LCC GE197 .C47 2020 | DDC 640.28/6—dc23
LC record available at https://lccn.loc.gov/2019019478

Columbia University Press books are printed on permanent and durable acid-free paper.
Printed in the United States of America

Cover design: Jordan Wannemacher
Cover image: ©iStockPhoto

For Robin,
my patient partner in this and other adventures.

CONTENTS

PREFACE

Environmentally minded Americans recognize that global climate change is the single most urgent ecological and political issue facing the planet. If you are like me, you want to live in a way that is consistent with your beliefs about climate change. We all know we need to reduce our carbon impacts. But we don't know by how much, and most environmental organizations don't give us a clue about what a sustainable carbon footprint would look like. The rate of global carbon emissions overwhelms us and makes individual action feel futile. And giving up carbon emissions entirely seems inconsistent with a contemporary, comfortable lifestyle in the developed world.

But most people share the basic ethical sense that it is wrong to make lifestyle choices that cause harm to other people. And we know that climate change will cause grievous harm to millions of people around the globe. It's easy to blame capitalism and large, impersonal oil and coal companies for climate change, but we can't ignore our own complicity in the fossil fuel economy when we burn gas to get to work, jet fuel to go on vacation, natural gas to heat and cook, and coal-generated electricity to light and cool our houses. As cartoonist Walt Kelly put it forty years ago, "We have met the enemy, and he is us!" Walt Kelly's poster for the first Earth Day celebration in 1970 showed woodland characters from his weekly Pogo comic strip despairing of the trash and detritus discarded in their forest.

For about ten years now, I have been trying to live on an individual carbon budget of about 4 tons per year, while still living well. I have gradually cut down on the biggest carbon emissions in my daily life, leaving some room for occasional carbon luxuries. Not incidentally, reducing your daily carbon emissions can save you money, too.

The idea for this book came to me in June of 2015 as I was preparing to give a talk at an environmental conference on defining and living within a defensible, quantified individual carbon budget. I wanted to total up my own carbon emissions for the year and compare them to the 4-ton budget I considered sustainable. But I was a little worried that I would expose myself as a hypocrite, since the previous twelve months had included international travel as well as the usual household emissions of living and working in the northeastern United States.

The easiest way to figure out my carbon tab was to go to an online carbon calculation website and see what my total was. The best carbon calculators let you enter exact numbers for the big-ticket items like natural gas or electricity use. So I went to one of the more thorough online carbon calculators, www.carbonfootprint.com, and plugged my travel and consumption figures in with the same foreboding I have when I do my taxes with a tax preparation program: did I blow my carbon budget that year?

I had those thoughts as I clicked open the calculator. No one would look at my previous year, or my life generally, and accuse me of being a self-denying ascetic. (One of my friends has even accused me of being a "fun hog.") That past year, I had skied powder in tall mountains, scuba'ed and snorkeled in translucent azure waters, lazed on tropical beaches, visited new countries and cultures, walked through some of the greatest cities on the planet, camped and paddled in the wilderness, and enjoyed many a fine meal and equally fine drink and music. After the annual Waterkeeper Alliance conference in June 2014, I rafted the rapids of the Youghiogheny River in western Pennsylvania. In July of that year I traveled to the volcanic Azores Islands in the Atlantic and climbed Pico, their highest peak; I also visited Spain and Portugal's ancient cities of Santiago de Compostela, Porto, and Lisbon. During the fall, I commuted by zippy motorcycle to my job lecturing in environmental law. I made a trip north to Vermont for the New England foliage season, piggybacking on an academic conference. During a winter-spring sabbatical, I split my time between our cozy house in the suburban woods in Rockland County, New York, and a ski cabin in the mountains. I attended a conference in Washington, DC.

In late winter, I traveled to Dakar, Senegal, to lecture on clean water regulation at a conference of West African environmental activists. Taking advantage of my sabbatical freedom, on the way home I visited the Caribbean islands of Guadeloupe, the U.S. Virgin Islands, the Turks and Caicos, and the Bahamas.

And yet I discovered, to my great relief and surprise when I went to carbonfootprint.com and plugged in my numbers, that I had done it all while keeping close to my 4-ton annual carbon budget.

The carbon footprint calculator asked for my auto travel, electricity usage and provider, air travel, and train travel for the year, in addition to some questions about eating, shopping, and recycling habits. My individual carbon footprint for the year, including all that international travel, those ski trips, heat, and utilities, was about 4.2 tons of carbon dioxide, including an estimate of the elusive "indirect impacts"—not far above my personal goal of 4 tons per year (though the current version of the calculator would assign a higher value to indirect impacts—those over which an individual has no direct control). This represented about 40 percent of the average U.S. individual direct carbon footprint.

How was this possible? To be sure, that year was not a typical year in my life or in the life of anyone with a full-time job. The flexibility of an academic schedule and a sabbatical semester freed me from many of the constraints of modern life. This flexibility allowed me to travel nearly carbon free on my sailboat from New York to the Azores, Spain, and Portugal in the summer of 2014, and to return with the boat via Dakar and the Caribbean in the spring of 2015. Our ski cabin in the Adirondack Mountains of northern New York is off the grid, solar powered, and heated with deadwood from the surrounding forest lands, and our home in Rockland County is also partially heated with wood from our downstate backyard. The motorcycle I commute on is electric powered and charged with solar panels on my roof at home. I pay a few pennies more per kilowatt at home for an all-wind-power electric supplier. I didn't own my own car, though I shared a 50-miles-per-gallon Prius with my wife, which we used for travel to the mountains for skiing. I took the train to Pittsburgh and DC for my conferences, so my only air travel for the year was one round trip between Spain and New York. My 4 tons or so of carbon footprint included this plane flight (the biggest single item), my share of the natural gas heat for our house, my share of the diesel we burned on our sailboat getting through calms and in and out of harbors, and my share of the gas we burned getting up to our ski cabin and back.

This past sabbatical year may not be typical. But for some five years now I have been living happily in suburbia, commuting to work, having fun on weekends, and traveling to interesting places on vacations—all while keeping close to my 4-ton carbon budget.

Now I would like to share my experiences.

The dual, mildly contradictory, premises of this book are that (1) we all share a personal ethical obligation to maintain a sustainable carbon footprint and that (2) life should be fun and fulfilling. This book will not try to convince anyone of the ecological, political, and humanitarian disasters that will arise from climate change, or that climate change is caused chiefly by the human activity of burning fossil fuels. There are plenty of excellent treatments of that subject. This book is for people who already recognize that global climate change is the single most urgent ecological and political issue facing the planet and who wish to adjust their own lifestyle to be consistent with their beliefs about climate change. In a word, this book is written for people who are committed environmentalists. This book is not for environmentally conscious 1-percenters, either; there are plenty of stories of celebrities spending millions of dollars to build a state-of-the-art, geothermally heated, solar panel–roofed, triple-insulated mansion with two electric Tesla roadsters in the garage, while they jet off to Kilimanjaro in their private jets to see the last snows before they melt with global warming. These are inspiring stories despite the mixed carbon footprint message, but those choices are far out of reach for most of us.

Nor is this book written for the hair-shirted ecohermits living full-time in an off-the-grid cabin with 40 acres of woodlot, home-grown vegetables, a flock of chickens, and a milk cow (*Mother Earth News* gives great advice to this group on a quarterly basis) or for their urban self-denying counterparts. *No Impact Man* by Colin Beavan tells an inspiring story of personal commitment, but it leaves most people feeling, "I really don't want to live like that," or worse, "What a self-righteous bore; his poor wife and children. Must be nice not having to go to work every day." Even Beavan didn't want to live more than one year with "no impact."

Rather, this book sets out to establish a sustainable new American dream, with a family living in a single-family house in the suburbs, raising children, commuting to work on a daily basis, and enriching their lives and having fun when they can—all while keeping their carbon impact to a sustainable level. The global poor, which unfortunately continues to mean most of the people on the planet, live sustainably because they have no choice. Sustainability has little meaning for the global superrich as long as they remain few enough in number to mute the outsized impacts of lifestyles that would not be remotely supportable for the masses.

I know, the reader might be thinking, "ocean sailing, ski cabin? Sounds pretty 1-percent to me. How are the rest of us supposed to have fun on your carbon budget?" There is no question that I am fortunate in life. Like most middle-class Americans, I am one of the global lottery winners. Twenty years ago I leapt at the chance to become a professor, trading a law firm income for the autonomy and flexibility that come with a career in legal academia and compensation that was less than from a private law practice but more than adequate. And it turns out that a low-carbon lifestyle will save you money, especially if you give up the big-ticket items like a large house, multiple automobiles, and regular air travel. Ocean sailing has a bum rap as a rich man's sport: buying and maintaining a used offshore-capable sailboat costs about as much as owning a recreational vehicle or a large sport-utility vehicle, but without the carbon footprint. And our ski cabin is quite modest—just one room.

This book is my attempt to articulate a philosophy of individual carbon sustainability in the developed world, as well as to document my own attempt to live—and live well—with a sustainable carbon footprint for one year. As I explain in chapter 3, I believe that an achievable and defensible individual direct carbon footprint for a middle-class American is about 4 tons of carbon dioxide per year, representing about 40 percent of the U.S. average direct carbon footprint. In between the substantive chapters I report on my own success in sticking to this carbon budget for one year while getting myself to work every day in the suburban New York communities where I live and work—and having fun on weekends and vacations.

If the world is going to reduce its carbon footprint to sustainable levels, someone has to start by showing that it can be done by individuals in the comfortable classes in the developed world, the source of most of the planet-destroying greenhouse gas emissions. If committed environmentalists don't start living sustainably now, the environmental movement at large has zero chance of convincing the rest of society to make the structural changes necessary to save the planet.

The technologies for low-carbon living have already been invented, they are available as mass consumer products, and they are available now. Some of them, like the Tesla roadsters or (for much less money) Zero electric motorcycles, are among the most exciting consumer products available on the market today, carbon spewing or carbon free. And many of the choices for a low-carbon lifestyle come with huge side benefits: improved health, improved self-image, sense of independence and self-reliance, and increased connection with family, life partner, and local communities.

So go ahead.
Live. Sustainably. Now.

LIVE SUSTAINABLY NOW

PROLOGUE

Climate March at the Crossroads

The People's Climate March in New York City in September 2014 held the promise of being a turning point in climate activism. Four hundred thousand people gathered in what was to date the largest march in favor of action to combat climate-altering greenhouse gas emissions. My wife Robin and I joined the march on a fine fall day. We ran into friends on the train down to the city and joined the throngs walking uptown from Grand Central Station to the staging grounds. Robin went as the planet, wearing an earth-skirt and carrying a polar ice cap umbrella. Many other marchers were similarly dressed for the occasion. I was a little less outlandish in my Waterkeeper Alliance T-shirt. Robin went to march with the earth scientists, while I joined the activists section, where Waterkeepers from around the country were assembled. The march was part protest, part festival, and part family reunion. My brothers and sisters from Waterkeeper had traveled from across the country and around the world to be part of this watershed moment, and it was great to get caught up with my friends on the front lines of environmental advocacy. Chants, slogans, and inspiring speeches filled the air.

Five years earlier, political action to address climate change had seemed hopeless. The organized environmental community had pinned its hopes on the new Obama administration and Democratic control of Congress to succeed in enacting climate change legislation to reduce U.S. greenhouse gas emissions. The big green political wing had bet on a bipartisan bill to put a cap on carbon emissions, with a trading system that would allow industries to buy and sell carbon emissions rights. But the market trading part of the bill was not attractive enough to win any Republican votes. And in the wake of the 2008 financial collapse, *market trading* was a dirty word on the left. The proposed Waxman-Markey bill was so timid that it dared not even include the word *climate* or *greenhouse* in its name; the bill's sponsors preferred to call it the "American Clean Energy and Security Act." The actual bill was over a thousand pages long and chock-full of special provisions and compromises that looked suspiciously like loopholes. Few environmentalists—much less the general public—actually understood what the bill hoped to accomplish, or how. The legislation never made it to the floor in the Senate, and President Obama calculated that universal health care would be a more attainable mark for the signature legislative initiative of his presidency.

Five years later, the Peoples Climate March seemed to be the answer to the calls by political analysts for a popularization of the climate fight. Both political scientist Theda Skocpol, in her postmortem study of the failure of the Waxman-Markey cap-and-trade bill, and journalist Naomi Klein, in her 2014 book *This Changes Everything*, had called for this sort of mass popular uprising to take the climate debate away from the beltway insiders and carry it out to the districts. Climate political action had turned away from the legislative arena and turned to specific projects, like the Keystone XL pipeline, that were easier for the public to understand.

The organizers of the People's Climate March wore the mantle of the civil rights marchers of the 1960s. Indeed, the lead speakers consciously included people of color and veterans of the civil rights struggle. But participants and observers of the march might have been left with a different impression: the feeling of the event was more one of celebration than uprising, with colorful costumes, floats, music, and drummers. The crowd was thrilled to take part in a historic event with thousands of like-minded people. The parade was lawfully permitted, the police were friendly and supportive, and the marchers generally behaved themselves. Civil disobedience was left to the more radical Occupy Climate protesters downtown. A politician watching the event would not have been left with the impression that this particular movement posed any threat to the established order or required accommodation with some climate version of the Voting Rights Act. An industry executive might likewise feel no threat to the established fossil fuel–powered economy; indeed, tallying up the diesel, gas, and jet fuel bought and burned to bring all these people together in New York City, the march was undoubtedly a net positive for the oil industry's bottom line.

And this unspoken reliance by the climate marchers on the very fossil fuel economy they purportedly sought to dismantle may continue to be the Achilles heel of the climate movement in its current form because, unlike the marchers in Selma or Montgomery, Alabama, actual victims of climate change were largely missing from the assembled climate activists. The civil rights marches were organized and carried out nearly entirely by African American victims of racial injustice in the United States. The Peoples Climate March set aside only two city blocks (65th to 66th Streets) for groups representing the climate victims—the "front lines of climate change"—and an overflowing twenty blocks for everyone else. These "frontline marchers" included people displaced by Hurricanes Katrina and Sandy, indigenous people from island nations threatened with inundation by sea level rise, and low-income communities that suffer disproportionately from environmental hazards. In keeping with the spirit of the march, the indigenous peoples wore traditional dress and the frontline community representatives carried banners and protest signs. But these "frontline" representatives were a small minority of the marchers.

In fact, based on the simple inexorable math, mere attendance at the Peoples Climate March pretty much established an individual as a perpetrator of climate injustice rather than a victim. While the climate marchers may have drawn inspiration from the 1960s civil rights marches, their lack of any true personal grievance risked less favorable comparisons to an earlier social movement, the temperance movement, that achieved the colossal political success and practical failure of prohibition. Climate activists take a risk to the extent that they base their movement, like the temperance movement, on validating values rather than alleviating true grievances. On the other hand, the ultimately successful abolition movement may provide an example of a successful values-based social movement. The question is: How can the climate movement avoid the mistakes of prohibition? And, as more voices recognize that the only effective response to climate change is fossil fuel abolition, the climate movement should be asking itself how it might mimic the long-term success of the slavery abolition movement.

The answer may lie in a level of personal sacrifice and commitment that the climate movement to date has been unwilling to demand. Abolitionism and temperance both expected abstention from movement activists. Teetotaling temperance activists did not actually give up anything they cared about. Abolitionists, on the other hand, did, as they boycotted slave-produced sugar and cotton. Are climate activists ready to give up things they care about?

1

CLIMATARIANISM

Our Personal Moral Obligation

In June 2015, I gave a talk on personal carbon sustainability at the Waterkeeper Alliance annual conference in Boulder, Colorado (in case you are wondering, I took the train from New York to reduce my carbon footprint). There were about twenty attendees at my session, all environmental leaders in their communities. At the start of the session, I asked the question "Who believes that all people have an ethical obligation to reduce their individual carbon footprint to a sustainable level, even if the people around them are not doing so?" Not surprisingly, given the audience, every hand went up. A few years earlier, in 2012, author and climate activist Bill McKibben had written an influential article in *Rolling Stone* explaining how the hard limits on carbon emissions required to meet a 2°C limit on global warming meant that most fossil fuel reserves would have to be left in the ground. I explained to the session that the 562-gigaton global budget for greenhouse gas emissions popularized by Bill McKibben's article implied a global limit of 2 tons per capita per year. I talked about what this meant in terms of individual carbon-generating activities like driving a car, heating your house, or taking a plane trip. At the end of the lecture, I asked the question again—whether all people have an individual ethical obligation to limit their carbon footprint to a sustainable level—and only two or three hands went up.

The overwhelming nature of the climate problem tends to do that to people, clouding their judgment on seemingly simple questions of personal ethics.

FIRST: DO NO HARM

It seems self-evident that we all have a moral duty to avoid emitting greenhouse gases. After all, it is a universal ethical principle that one must not act in a manner that causes serious harm to other people. Greenhouse gas emissions will cause climate change that will cause grievous harm to individuals inundated by rising seas, starved by agricultural collapse in desertifying regions, and victimized by the political instability arising from climate refugees. Even the huge populations that may not qualify as climate refugees will be affected. Even among people who are not displaced by climate change, rapid deterioration and alteration of familiar natural ecosystems will cause psychological stress and loss of a sense of well-being. This sort of environmental psychological stress has already been identified in communities of Appalachia where the natural landscape has been torn up by mountaintop removal.

Climate change thus clearly invokes the avoidance-of-harm principle. The ethics of greenhouse gas emissions also invokes a variation of the golden rule, "Do unto others as you would have others do unto you." If you would have the rest of the world avoid excessive greenhouse gas emissions, it behooves you to avoid excessive greenhouse emissions yourself.

CONSEQUENTIALIST ETHICS AND YOUR CARBON FOOTPRINT

This simple application of the golden rule should settle the issue. It does for me. But as with most simple statements, there is a lively debate about how this rule applies to our particular circumstances. Believe me, as a lawyer and law professor, I know all about getting into lively debates about the application of simply stated rules to particular circumstances.

So professional philosophers have joined the debate about climate ethics. And where a simple ethical rule seems to conflict with the lifestyle choices of mainstream society, it seems we ought to pay some attention to the people—largely academics—who make a living thinking about ethical issues. We might find some help in defining exactly how a general ethical principle might apply to a thorny problem like climate ethics.

Philosophers disagree about the existence of a personal ethical obligation to reduce one's greenhouse gas footprint. Some philosophers have argued that we have no ethical duty to people who don't even exist yet, so that the effects of changing climate on future generations do not give rise to ethical obligations on the part of the current generation.

Another argument against a personal ethical duty distinguishes between ethical obligations at a national or societal level and those at a personal level. This is the "inconsequentialist" argument against individual ethical responsibility, also called the "causation" argument. It reasons that since any individual's contribution to global greenhouse gas emissions is such a small portion of the problem, and since even completely eliminating any single individual's greenhouse gas emissions will not stop harmful climate change from occurring, individual emissions are "inconsequential" and cannot be said to "cause" harm to anyone. One academic philosopher even claims that there is nothing morally questionable about spending an afternoon joyriding in a gas-guzzling sport utility vehicle (SUV) just for the fun of it.

Each of these objections has been answered. There is no philosophical justification for ignoring the rights and interests of persons not yet born; they should enjoy the same basic human right to an environment that will support them as does the current generation. In any event, individuals now in existence have already suffered, and will suffer during their lifetimes, various harms arising from greenhouse gas emissions.

The causation issue is more difficult. Intuitively, people think that "it doesn't matter if I reduce my greenhouse gas emissions; since everyone else is still spewing carbon dioxide into the air, global warming will still happen, and it is pointless to deny myself any pleasure in life when it will make no difference." This seems like an argument against personal action, even for those who advocate for strong government action to limit collective emissions. In fact, this argument captures the classic problem of collective action, the "tragedy of the commons" that is the basis for environmental regulation generally. It is also sometimes referred to as the "free rider" problem—the tendency of individuals to take a "free ride" on the collective sacrifice of others. (I will discuss this problem further in chapters 2 and 6.)

But the causation argument assumes that climate change is a problem with a threshold impact; that is, greenhouse gas emissions are going to cross a critical threshold with or without my individual contribution, so my contribution doesn't matter. In fact, climate impacts are incremental, so all emissions, no matter how small, contribute to the gravity and extent of the overall harm. One climate theoretician has come up with a rough equivalence: four days of excess emissions in the developed world now in order to enjoy a richer life roughly equal one day of a ruined life of someone living in a climate impact zone at the end of the century.

The ethics of personal greenhouse gas emissions is further complicated by the impossibility of taking an uncompromisingly perfect position in this debate. We all emit carbon dioxide every time we exhale. In fact, the combined carbon dioxide exhalations of the world's population constitute an impressive 3 to 4 percent of current human greenhouse gas emissions. Of course, the carbon dioxide we breathe out was all recycled by photosynthesis in growing the food we eat, so these emissions don't add to global carbon dioxide concentrations. But the global food production system does. Even of you are on a 100 percent organic vegan diet, it is pretty much impossible to buy food that is produced without adding to climate change. In other words, everyone in the developed world who is eating contributes to global warming that will cause grievous harm to climate change victims. But of course, any argument that leads to the conclusion that eating is unethical is not going to be widely accepted.

So it can't be that all greenhouse gas emissions are unethical, only those emissions in excess of . . . what? Suddenly, the problem of the ethics of greenhouse gas emissions has turned from a question of right or wrong to a question of "how much is too much?" This seems like a question of societal distributive justice—what is a fair way to distribute a limited resource?—rather than of personal ethical responsibility. Is it unethical to take more than your fair share of the chocolate chip cookies on the dessert plate, or is it just bad manners? And what is a fair share of the dessert cookies anyway? Should larger, hungrier individuals be entitled to more of the cookies because they are larger and hungrier? Or should the cookies be distributed equally to everyone at the table? Or should the undernourished starving person outside the house be given a cookie before the well-fed people inside help themselves to dessert?

If you think I am setting up an allegory about the developed nations and the developing world with this example, you are right. Do I have a satisfactory answer to this problem of distributive justice? Nope. But there is a compelling argument based on accepted principles of distributive justice that the developed world, which benefited from two centuries of unrestrained emissions, can't possibly claim entitlement to more than a per capita allocation of the remaining greenhouse gas increment. Distributive justice usually looks to bring the least well-off people within a given economic system up to an acceptable standard of living. When it comes to carbon emissions, the economic system is global, not national, since the atmospheric climate system and its impacts are global, not national. The developed nations are in a poor position to argue that they should be allowed to continue to produce above-average per capita emissions to maintain a standard of living well above the global average. But an equal global distribution of permissible carbon emissions takes us back to an austere 2 tons per person. This maximum fairness solution is not likely to be achieved in our existing global politico-economic system, which, for better or worse, allocates limited resources through global markets according to wealth and income. If someone has an allocation system that works better for the greatest number of people, let me know.

In a 2015 book entitled *Climate Change and Individual Responsibility*, a Belgian ethics philosopher, Wouter Peeters, and his colleagues have tried to answer the question of what ethics demands in terms of individual carbon footprints. They debunk each of the philosophical arguments against an individual moral responsibility to limit emissions, noting that there is no justification to discount the basic human rights of future persons or the human rights of existing persons who will suffer future impacts. They emphasize the incremental effect of greenhouse gas emissions on climate change, citing an analysis that shows that one typical American life, with its attendant greenhouse gas emissions, will cause two climate-related deaths in the developing world due to climate impacts.

Peeters et al. also consider the various mechanisms of moral disengagement that well-meaning individuals in the developed world invoke to avoid accepting responsibility for their carbon footprint. We all focus on the government's failure to act rather than our own actions; that is, we define the climate problem in terms of government's failure to take measures to limit emissions rather than our own active (and more blameworthy) decisions to engage in activities that cause emissions.

People are receptive to the arguments that climate change is not caused by humans, however flimsily based, because these arguments deny that we can foresee the climate harms that result from our individual conduct. (Of course, this is no excuse for those who accept the conclusion that human emissions are to blame for climate change.) If the harms are not foreseeable, we can't be guilty of causing harm intentionally and knowingly. We rely on the commonplace nature of the climate-changing conduct in our society. Everyone (so it seems) drives a car and rides on airplanes, and if everyone is doing it, it can't be morally reprehensible—though we conveniently ignore the fact that, globally speaking, hardly anyone drives a car or rides on airplanes. We find justifications for our own emissions: surely this trip to the mountains, or to a unique landscape on the planet, is a worthy pursuit.

Peeters et al. see this reasoning as a form of moral disengagement—ethical avoidance rather than ethical reasoning. Significantly, this disengagement puts responsibility on the government rather than the individual. Many of us resolve the cognitive dissonance of burning fossil fuels in our daily lives despite their global warming impact by putting our energies into activism for government action to limit emissions. But, as Peeters et al. point out, any particular individual political action is just as inconsequential in terms of preventing climate harms as an individual's carbon footprint reduction. Political action must be widespread and collective to be effective—and widespread individual actions can also significantly reduce climate impacts. Peeters et al. explain:

> Individual agents cooperate in formal and informal polycentric systems . . . which can elicit major benefits at multiple scales in terms of emissions reductions, information dissemination and individual value change. . . . The statist approach underestimates the potential, which can be achieved on individual and 'non state' collective levels of action.

Drawing on the work of other climate ethicists, *Climate Change and Individual Responsibility* resolves the question of "how big does a carbon footprint need to be before it becomes unethical?" by relating the ethics of carbon emissions to the idea of subsistence: the act of maintaining oneself at a minimum level. Peeters and colleagues argue that one has no moral obligation to sacrifice one's own basic human right to subsistence in order to protect another's right to subsistence; this means that emissions that are needed for subsistence are morally OK. But any individual emissions in excess of those necessary for subsistence—so-called luxury emissions—become ethically "reprehensible" (meaning, simply, that these actions are not ethical). Peeters et al. conclude that these emissions in excess of subsistence can be said to cause climate harm to other persons in a way that subsistence emissions do not. It might be more accurate to say that subsistence emissions also cause climate harms but that they are justifiable harms because they are necessary to preserve life, and therefore morally acceptable.

The authors thus give one answer to the question of personal ethical responsibility for greenhouse gas emissions: it's OK to keep eating and heat your house, but any emissions in excess of the amount necessary for your subsistence is morally wrong.

There are two practical problems with this conclusion. One, which Peeters et al. acknowledge, is that the definition of subsistence emissions is highly subjective and dependent on cultural context. What might be considered subsistence in the United States would be very different from what would be considered subsistence in the Sudan. As an example of an unjustifiable "nonsubsistence" carbon emission, Peeters et al. suggest that setting your air conditioner to 23°C (about 73°F) rather than 28°C (about 82°F) would exceed subsistence levels of air conditioning and would be morally reprehensible. But this begs the question whether any air conditioning is necessary for subsistence when the majority of the world's population live in tropical regions without any air conditioning. The Bangladeshi family flooded from their home by rising seas will likely have lived without air conditioning most of their life, and they might not be convinced that even 28°C air conditioning is essential to the subsistence of the middle-class Western family.

So the subsistence test may, in some sense, be insufficiently austere to be internally consistent. On the other hand, it may also be too austere to gain widespread acceptance as a cultural ethical norm. In short, the subsistence principle does not allow for fun, for any (carbon-emitting) human activities engaged in for pure pleasure and life enrichment. A moral philosophy that limits everyone to the bare essentials to maintain life, with no luxuries whatsoever, will not likely have much more appeal than one that treats the act of eating as a moral wrong.

A moral philosophy that allows one to eat and breathe, but not much else, isn't going anywhere. As Tennyson's aging Ulysses put it, "I cannot rest from travel. . . . / As tho' to breathe were life! Life piled on life / Were all too little." Sure, a subsistence-based ethical limit on carbon emissions makes it clear that joyriding in a gas-guzzling SUV on a Sunday afternoon is ethically wrong. But driving a fuel-efficient hybrid 10 miles to a wilderness park to go for a walk in an awe-inspiring natural forest is equally wrong: that nature walk is not necessary for subsistence, and driving there contributes to climate change impacts that have no threshold.

OTHER ETHICAL APPROACHES TO YOUR CARBON FOOTPRINT

These ethical arguments, which are based on the avoidance of harm to other people and the consequent problems of defining excess in terms of causation, live under the rubric of consequentialist ethics. This utilitarian school of ethics develops and applies ethical principles based on their ability to advance human good in terms of well-being and happiness and to avoid human harms such as sickness, displacement, and death. Actions and decisions are not ethical or unethical in the abstract; rather, the morality of any given decision depends on its consequences.

Traditional Western ethical thought focused exclusively on human well-being and harms. However, the "deep ecology" movement of the latter part of the twentieth century recognized that natural systems have their own intrinsic value, independent of their utility to human well-being. Deep ecology is not inconsistent with a causation approach to ethics; it just adds ecosystem consequences to the consequences that must be considered in evaluating the morality of a given action. The argument for avoiding action that contributes to global warming gets much stronger when ecosystem integrity is added to human well-being as a consequentialist good.

Other ethical systems exist, of course. There are rule-based ethical systems; ethicists call this branch of ethics "deontological ethics." There is also "virtue ethics," which traces its roots to the Greek philosopher Aristotle and focuses on neither acts nor consequences as much as the consistency of human conduct with the virtues necessary for human flourishing.

Rule-based ethical systems look at the ethics of individual acts, without regard for their consequences. The Ten Commandments of the Bible are an example of rule-based ethics. Rule-based ethical systems do not have to be religiously generated, and many rule-based ethical systems draw on moral intuitions and traditions of natural law. But these traditions have not always included considerations of environmental stewardship. And moral intuitions can be subjective and manipulative, so that reliance on moral intuition may not provide a single answer to an ethical dilemma like personal greenhouse gas emissions.

One example of a "rule" of environmental ethics comes from naturalist Aldo Leopold's *Sand County Almanac*. Leopold wrote this series of essays on the beauties of the natural environment of rural Wisconsin during the 1940s. The essays were part of the philosophical inspiration for the environmental movement of the latter half of the twentieth century. In one essay, "Thinking like a Mountain," Leopold laments his own participation in the misguided slaughter of a she-wolf and her pups. Realizing that destruction of a predator species does not ultimately mean a healthier deer herd for hunters, and attributing feelings of sorrow to the landscape itself, Leopold intuited an ethical rule calling for respect for the natural balance of ecological systems: "A thing is right when it tends to preserve the integrity, stability, and beauty of the biotic community. It is wrong when it tends otherwise."

Leopold called this rule the "land ethic." Recognition of this ethic carries with it the ethical responsibility to avoid activities that destroy the integrity, stability, and beauty of the biotic community. It then follows that individuals must reduce personal greenhouse gas emissions that contribute to climate change, as climate change is the ultimate disruption to the integrity, stability, and beauty of the global biotic community.

Religious proscriptions are also a form of rule-based, or deontological, ethics. The Ten Commandments are the paradigm for an ethics based on rules derived from religious authority. And world religious leaders are now beginning to recognize the theological grounds for a proscription against harming the global climate system. The most striking development, of course, is Pope Francis's encyclical on the environment and climate change, issued on June 18, 2015, the *Laudato Si'*. Pope Francis wrote:

> The creation accounts in the book of Genesis contain, in their own symbolic and narrative language, profound teachings about human existence and its historical reality. They suggest that human life is grounded in three fundamental and closely intertwined relationships: with God, with our neighbor and with the earth itself. . . . A very solid scientific consensus indicates that we are presently witnessing a disturbing warming of the climatic system. In recent decades this warming has been accompanied by a constant rise in the sea level and, it would appear, by an increase in extreme weather events, even if a scientifically determinable cause cannot be assigned to each particular event. Humanity is called to recognize the need for changes in lifestyle, production, and consumption to combat this warming or at least the human causes that produce or aggravate it.

Pope Francis has thus brought Catholic orthodoxy into the creation-care movement of Christianity, which similarly holds that humans are stewards of God's creation and have a responsibility to preserve the global ecosystem intact. This recognition of stewardship obligations stands in contrast to the so-called dominion theory of Judeo-Christian environmental thought, which relies on the biblical injunction to humans to "be fruitful and multiply, and replenish the earth, and subdue it: and have dominion over the fish of the sea, and over fowl of the air, and over every living thing that moveth upon the earth." Following Pope Francis's lead, a group of Islamic religious and environmental leaders has issued its own call to phase out nonrenewable energy and stop greenhouse gas emissions no later than 2050.

FIGHTING CLIMATE CHANGE THROUGH CONSPICUOUS NONCONSUMPTION

Conspicuous consumption describes the consumer practice of making purchases as a way of communicating social status as much as for the practical utility of the item. This practice leads to purchases of cars more powerful than necessary for transportation purposes and houses much larger than necessary for shelter. Product advertising takes advantage of the basic human desire to display social acceptance and leadership—and convinces people that the cars they drive, the foods they eat, and the places they travel to will make them more attractive socially. Climate-conscious consumers can buck this trend by engaging in "conspicuous nonconsumption." When you truly understand that your consumer choices damage the global climate balance, the choice not to consume fossil fuels can communicate your status as a climate leader. Choosing to reduce your carbon footprint and letting your friends and professional circle know about it help communicate the urgency of the climate crisis as well as the availability of more climate-friendly approaches to life. It may seem like a small start—but by modeling climate-responsible behavior you help promote social acceptance of low-carbon choices. In the words of Mahatma Gandhi, "Be the change you wish to see in the world." *Being the Change* is the title of a book by climate scientist Peter Kalmus, who also argues for reducing one's individual footprint to address climate change.

If you are not convinced by the pope's ethical arguments, there are at least two other, more practical arguments for an individual response to climate change. First, even though reducing your individual climate footprint may not make a measurable dent in the global temperature increase, your choice to reduce emissions helps communicate your own sense of the gravity of climate change to others. A 2016 study on the effectiveness of climate communications by climate scientists revealed that those scientists who had taken actions to reduce their own carbon footprints were seen by laypeople as more credible than those who had not taken individual footprint reduction measures. Even if you are not a climate scientist, your actions as an individual communicate an awareness of the seriousness of your concern about climate—and that helps move the political consensus toward necessary collective action to address climate change.

Second, taking individual action to reduce your climate footprint will help reduce your own "climate stress." The gravity of the threat of climate change, combined with the enormity of the social and economic changes needed to address the problem, tends to overwhelm people emotionally—and this stress is a real problem. Taking personal action helps relieve it. For many individuals concerned about climate change, personal action translates into political action—marches and direct action blocking fossil fuel pipelines. But not everyone who cares deeply about climate change has the personality to engage in acts of civil disobedience. Taking individual action to reduce your footprint is a way to do what you can to address climate change without risking jail time. Taking action to live according to your deeply felt belief about climate change will reduce the cognitive dissonance of living in the developed world with its unsustainable energy economy. It is also a way to take an overwhelming problem—the grave global consequences of climate change—and convert it into meaningful action with measurable results. Yes, your personal carbon reduction may be small in the context of global greenhouse gas emissions, but you can measure it and demonstrate it and be certain that you have contributed to progress on the global task of reducing emissions. And, yes, this is a form of ethics—the Aristotelian system of "virtue ethics," which is based on making ethical choices that contribute to the well-being of individuals who live according to their ethical virtues, not because of the practical consequences (utilitarian ethics) or the adherence to rule-based ethics (deontologism).

So where does all this leave us? Is there a personal ethical responsibility to limit one's greenhouse gas emissions? And if so, what level of greenhouse gas emissions beyond simply breathing is morally defensible? I think the answer to the first question is a clear "yes" under either a utilitarian or ecology ethics approach. All greenhouse gas emissions contribute incrementally to climate change, and climate change will cause grievous harm to people and the ecosystems that support people both materially and spiritually.

For anyone who has accepted the basic science and the urgency of addressing climate change, adopting a personal ethic of limiting one's carbon footprint to an acceptable level helps avoid the cognitive dissonance and sense of helplessness associated with taking the climate problem seriously. For those who are environmental leaders in their community, the personal imperative to limit one's carbon footprint is even stronger, for these leaders set an example for other environmentally minded citizens in their community; they are essentially a force multiplier for their personal carbon reductions.

The answer to the second question—what is an ethically acceptable carbon footprint in the developed world?—remains more difficult. In the long run, a simple answer might be that all burning of fossil fuels must cease, and an acceptable carbon footprint is one that involves no fossil fuels. At the rate the world is emitting greenhouse gases now, we will have to ban fossil fuels globally by around 2040 if we are to avoid disastrous climate change in excess of 2°C. An existence completely free of fossil fuels is not practicable right now in the United States for someone with a grid-connected house and regular employment. Nevertheless, everyone's goal in life should be to get to zero fossil fuel use by 2040 as a matter of global necessity. This book will try to look at ways to plan for zero fossil fuel use in that timeframe.

But what is a defensible footprint now, at the end of the second decade of the twenty-first century? Peeters and colleagues' subsistence-based limit on carbon emissions is too strict to be practicable and is as likely to draw a significant backlash against environmentalists as fun-hating, preachy ecofreaks. An equal per capita allocation, as may be demanded by strict application of the golden rule—"emit greenhouse gases as you would have every other individual on the planet emit greenhouse gases"—likewise leads to an emissions limit so austere as to preclude much fun in life.

A slightly modified version of the golden rule might help provide a workable answer. If we accept that global economic inequality is never going to disappear entirely, and that greenhouse gas emissions (like other scarce resources) could be distributed generally according to wealth and income, then a workable variation on the golden rule might be to "emit greenhouse gases at a level that, if emitted by everyone else on the planet at my level of wealth and income, would not exceed sustainable limits on greenhouse gas emissions necessary to avoid calamitous climate change." This approach allows individuals to find a limit that avoids causing climate impacts while recognizing that global inequality is not going away soon and that asking someone in a developed country to live in the midst of a society of excess with the carbon footprint of a sub-Saharan African will never achieve social acceptance. And it allows individuals, within the constraints of sustainability, to choose carbon-emitting "fun" activities as long as their overall footprint remains sustainable.

Of course, this begs the question: "What is a sustainable carbon footprint for a middle-class person in the United States?" Chapter 4 will discuss the concept of sustainability and the planetary limits on cumulative carbon emissions, and chapter 5 will try to translate sustainability principles into a number.

REFERENCES

Almassi, Ben. "Climate Change and the Ethics of Individual Emissions: A Response to Sinnott-Armstrong." *Perspectives: International Postgraduate Journal of Philosophy* 4, no. 1 (2012): 4–21.

Attari, Shahzeen, David H. Krantz, and Elke U. Weber. "Statements About Climate Researchers' Carbon Footprints Affect Their Credibility and the Impact of Their Advice." *Climatic Change* 138, nos. 1–2 (September 2016): 325–338.

Baer, Paul. "Equity, Greenhouse Gas Emissions, and Global Common Resources." In *Climate Change Policy: A Survey*, ed. Stephen Schneider et al., 393–408. Washington, DC: Island Press, 2002.

Brennen, Andrew, and Yeuk-Sze Lo. "Environmental Ethics." In *Stanford Encyclopedia of Philosophy*, ed. Edward N. Zalta (Winter 2016). https://plato.stanford.edu/archives/win2016/entries/ethics-environmental/.

Brown, David, et al. *White Paper on the Ethical Dimensions of Climate Change*. Rock Ethics Institute at Penn State University, 2006. http://rockethics.psu.edu/documents/whitepapers/edccwhitepaper.pdf.

Callicott, J. Baird. *In Defense of the Land Ethic: Essays in Environmental Philosophy*. New York: State University of New York Press, 1989.

Dernbach, John C., and Donald A. Brown. "The Ethical Responsibility to Reduce Energy Consumption." *Hofstra Law Review* 37, no. 4 (2009): 985–1006.

Hall, Edith. *Aristotle's Way: How Ancient Wisdom Can Change Your Life*. New York: Penguin, 2019.

Peeters, Wouter, et al. *Climate Change and Individual Responsibility: Agency, Moral Disengagement, and the Motivational Gap*. London: Palgrave Macmillan, 2015.

CARBON DIARY

September 2015

I teach environmental law at Pace Law School in White Plains, New York, and I also run the Environmental Litigation Clinic there, where we train law students by letting them act as attorneys for the nonprofit environmental advocacy group Riverkeeper. As an academic, the cycle of my year begins each September with the resumption of classes and the return of regular work obligations. September is also an easy month to lead a low-carbon life in the Northeast: it's usually not so hot that you would need air conditioning, nor so cold to require heat, so it's easy to keep your household energy consumption down. September this year was unusually hot, so we kept the fan running in our bedroom every night.

Between my class schedule and regular student case management meetings, I am going to have to be on campus every day of the week, so I won't be able to allow myself the energy-saving luxury of telecommuting from home one day per week this semester. We live in West Nyack, New York, about 15 miles by road and across the Hudson River from my office. For about seventeen years now, I have been commuting to work a couple of days a week in good weather by paddling a kayak 3 miles across the Hudson River from Nyack to Tarrytown and then bicycling the 8 miles from Tarrytown to White Plains. It's a great way to start and end the day, and it keeps me in shape. It is a rare day when I look up at the traffic on the Tappan Zee Bridge from my kayak and think, "Gee, I wish I were up there in a car instead of here in my kayak this morning." I do it as much for the pleasure of starting the day with some outdoor exercise as for the energy and carbon savings. But the paddle takes time and can be dicey in rainy weather or high winds, so I am lucky if I can make my paddle-and-pedal

The author kayaking across the Hudson River to work.
Photo by Frank Becerra Jr./*Journal News*, courtesy of PARS International Corp.

commute work two days a week. The rest of the time I commute the 15 miles to work by riding an electric motorcycle or, in bad weather, taking the commuter bus that runs from West Nyack to downtown White Plains. It was a dry September this year, so I was able to kayak or motorcycle just about every day of the month, keeping the carbon impacts of my regular commute close to zero. When Robin was away during the last week of the month, I went out to an open mic session at a local pub one evening—and discovered that I could indeed ride my motorcycle with a guitar strapped to my back.

With the fine weather this September, we had plenty of opportunities for outdoor weekend getaways. Our son Justin and his girlfriend Danielle came to visit over Labor Day weekend, and we took them down the Hudson River in our sailboat from Nyack to New York Harbor and briefly out into the Atlantic Ocean past the Verazzano Narrows Bridge. Touring New York Harbor by boat is always a treat, and Danielle had never seen lower Manhattan, the Statue of Liberty, and Ellis Island before. We have made this trip many times, and we have learned that to go somewhere on the Hudson in a sailboat you need to either time your travel for the favorable tides or burn a lot of diesel to run the engine against the tide. We were lucky with the wind on the way down the river Saturday night, and we sailed all the way from Nyack to our overnight anchorage near the George Washington Bridge in Englewood. We then caught the start of the ebb tide early in the morning for our run out through New York Harbor and into the open ocean. We turned around at Coney Island just as the flood tide started running up the river. We hit a dead spot for about an hour and a half in New York Harbor and ran the engine to keep our tide until the sea breeze picked up and carried us all the way back north to Nyack.

The weekend after Labor Day we drove our daughter Beryl back to college in Amherst, Massachusetts, and then drove on up to our cabin in North River, New York, in the southern Adirondack Mountains. Our cabin is a special place for us, nestled deep in a stand of red pines and 30 acres of conifers in the north woods. A cross-country ski trail, part of the Garnet Hill Lodge resort, runs through the property. The cabin is off the grid, heated with a small wood stove. It has one open room with sleeping lofts. Solar panels charge some marine batteries that power a 12-volt electrical system to run the lights and the fan in the composting toilet and to charge our phones and tablets.

We spent the weekend sawing and chopping firewood for the coming winter, and we made time for hikes to Balm of Gilead Mountain on Saturday and OK Slip Falls on Sunday. Since Monday classes were cancelled for Rosh Hashanah, we stayed over at the cabin Sunday night, went for a delightful run in the cool of the mountain morning, and drove back downstate. On September 19, I led an orientation hike at Bear Mountain State Park for incoming environmental law students at Pace, and the following Sunday I took a colleague and his family for a pleasant sail on the Hudson River.

We enjoyed some urban attractions during the month as well, so our carbon footprint for the month includes driving to lower Manhattan for a friend's birthday party at a tango dance studio in Chelsea.

HOW I KEPT TRACK OF MY DIRECT CARBON FOOTPRINT

There is no universal definition of an individual's direct carbon footprint—those things an individual has direct control over. Your total footprint also includes an indirect component—those carbon emissions that result from community, government, and business activities not under an individual's control and the difficult-to-quantify life cycle impacts of the production of consumer goods. Clearly, buying and burning gasoline or natural gas are in an individual's direct footprint, as should be other energy consumption choices that necessarily involve the burning of fossil fuels, such as buying electricity generated by fossil fuel power plants and buying tickets to ride on fossil fuel–powered transport such as aircraft, buses, and trains.

For my personal direct carbon footprint budget, I kept detailed track of these fossil fuel–powered activities, which were included in online calculators at the time. I added the impacts of eating beef or lamb, foods that (due to methane production) necessarily involve substantial greenhouse gas emissions. To keep it simple, I did not try to calculate the impacts of producing the consumer goods I purchased or other foods, since these life cycle impacts get to be very uncertain because of the variations in energy sources used for production and transportation. I included work-related travel (since for the most part I have a choice in that), but I did not include any share of work-related energy consumption at my place of employment. These emissions can fairly be characterized as part of my indirect impacts.

A general rule of thumb is that an individual's indirect carbon footprint is roughly equal to his or her direct impact. Updated online calculators now provide more detailed information on the carbon footprint of consumer goods and services purchased than was available when I kept this diary in 2015 and 2016. A sample and explanation of my calculations are included as an appendix to this book.

My total carbon footprint for September was 405 pounds of carbon dioxide equivalent—well within a twelve-month, 4-ton carbon budget. Since I have a 100 percent renewable energy contract with my utility, I get to count my electricity use as zero carbon. So the electric motorcycle commute is mostly carbon free except for the top-off charging I do with an extension cord out my office window (I also charge the motorcycle at home with solar panels on my roof, when it is sunny enough). The biggest carbon impact—196 pounds—was for the natural gas used for cooking and for the water heater in our house. The other big carbon ticket item was the gas used to drive to Amherst and North River and back, but at least that was split with other people.

2

WHY BOTH INDIVIDUAL ACTION AND COLLECTIVE POLICY WILL BE NEEDED TO ADDRESS CLIMATE CHANGE

By arguing for individual climate ethics, I do not mean to suggest that individual emissions reductions, even if widely adopted, will be sufficient to address global climate change. Reducing your own footprint gives the personal satisfaction of knowing that global carbon emissions have been reduced measurably—and in units that can be measured in tons, as we will see in the next chapter. But an effective response to climate disruption will also require some combination of regulations on emissions, taxes or fees on emissions, and subsidies for development and implementation of fossil fuel alternatives.

It's easy to get bogged down in the details of the policy choices for responding to climate change, and climate advocates who are otherwise allies get into bitter intramural battles over which policy is the best. Some argue for putting a price on carbon in the form of a tax or tradable emissions rights, based on economic efficiency and market incentives for innovation; others reject these market-based solutions as unfair to the working people who pay a larger part of their income on fossil-based energy, and they argue for government regulation and alternative energy subsidies as the only workable solution. I have my own opinions, but I will not take sides in this book. Any climate policy is better than none, and achieving some kind of policy will take concerted political effort. The box at the end of this chapter briefly explains different policy approaches.

But while government action will be necessary to address climate change, it may not be sufficient. Not all government mandates become incorporated into our social and economic systems as hoped. For example, prohibition was a great political success but also a monumental political failure. This chapter will address the relationship between top-down government mandates and bottom-up grassroots change in making effective policy. It will argue that individuals have a role to play in achieving the culture change necessary to make government policies effective.

THE COLLECTIVE ACTION PROBLEM: "FREE RIDERS" AND INDIVIDUAL CLIMATE REDUCTIONS

Whatever your policy preference, some form of government policy will be necessary to address climate change. Significant individual reductions can't address the problem of those individuals who don't accept climate science or the necessity of reducing greenhouse gas emissions, or those who accept the science but are happy to let others make the reductions while continuing to produce high emissions themselves—the so-called "free riders."

The inconsequentialist argument against individual climate action, discussed in the last chapter, is based on the notion that no matter how widely adopted, individual climate reductions will never be enough to prevent the worst impacts of climate change. No matter how many people act voluntarily to reduce their carbon footprint, there will be many more people and companies who take advantage of the lack of regulation to continue burning fossil fuels. These unrestrained people will be the "free riders." They selfishly increase their own emissions to take up the slack provided by more altruistic citizens.

Many climate advocates argue that it is better for individuals to press for the needed government regulations than to engage in ineffectual acts of self-sacrifice. Ironically, many of the collective actions promoted by climate advocacy organizations, such as fossil fuel divestment and support for candidates who refuse to take fossil fuel industry donations, also suffer from barriers to effectiveness. These tactics, like individual carbon reductions, cannot address the climate problem unless they are very widely adopted. Divestment will not put oil companies out of business as long as some people with the money to do so keep buying their products and their securities. Politicians will keep accepting fossil fuel industry donations as long as, in their estimation, more people will vote for them because of the campaign advertisements that money will buy than will refuse to vote for them on climate principles.

Indeed, the act of voting itself poses a "free rider" problem: many people don't bother to vote because it is extremely unlikely that their vote will make a difference in the result of any election. Nonvoters take a "free ride" on the efforts of those who do vote, since the nonvoters rely on others who share their views to elect their representatives. Economists actually see the fact that people vote at all as a collective action paradox: the chance that an individual vote will matter is so small that it will never outweigh the time cost of going to the voting booth. Yet people vote because they believe in participating. And you can reduce your individual carbon footprint because you believe in participating in climate solutions.

Because of this "free rider" problem for any form of collective climate action, some form of government policy response will be necessary to bring all of our society along into a low-carbon economy consistent with a stable climate. Anyone committed to addressing climate change must take part in pushing for the necessary government response. This means, at a minimum, voting for candidates who make aggressive climate action part of their platform. But it also means raising the political profile of the climate issue. You can do this by calling and writing your representatives in Congress to demand action on climate legislation, writing letters to newspapers, supporting climate action organizations, and participating in climate marches. Above all, you must be unafraid to talk climate with your friends.

But government action, while necessary, may by itself not be capable of accomplishing the fundamental changes needed to address climate change. The assumption of climate policy activists is that with enough marches, calls, and letter-writing campaigns, the right representatives will be elected, other politicians will see the light, and Congress will enact laws limiting greenhouse gas emissions. Problem solved! But, as a lifelong student, teacher, and practitioner of our system of law and regulation, I can tell you that government regulation by itself is often not sufficient to achieve the kinds of deep social and economic changes necessary to wean Americans from a fossil fuel–powered economy and lifestyle. Examples such as prohibition and the school desegregation decision are cautionary tales about government policy that got ahead of social change and failed.

Even the abolition of slavery took a century to go from policy to practice throughout the United States.

ABSTENTION AND THE LIMITS OF POLICY AS CATALYST TO CULTURE CHANGE

Boycotting products and abstaining from using products have played a key role in successful political and social change movements. Politically successful movements that employed boycotts and abstention included the abolition movement, the temperance movement, and the civil rights movement.

Many abolitionists demanded abstinence from the economic evil they sought to undo. For the most part, nineteenth-century American abolitionists would not eat sugar or purchase cotton products produced by slave labor. They substituted honey and maple syrup for sugar, and linen and broadcloth for cotton fabrics. Regional abolitionist society meetings included resolutions that required not using the produce of slavery, such as this February 17, 1836, resolution of the Vermont Anti-Slavery Society: "Resolved, That by consuming the produce of the labor of slaves, we are directly sustaining the iniquitous system of slavery; and that therefore, as abolitionists, we are called upon to abstain from using such articles as are believed to come to us through a polluted channel." The pages of radical abolitionist William Lloyd Garrison's newspaper, *The Liberator*, are full of notices for stores carrying "free goods."

Similarly, the temperance movement famously demanded abstention from alcoholic beverages as a mark of devotion and membership in the movement. "Lips that touch liquor shall never touch mine" became the motto of the Women's Temperance Union. And while prohibition is now viewed as a monumental policy failure, it enjoyed monumental political success in its time. Few other movements can claim to have passed a constitutional amendment enshrining their cause; abolition and women's suffrage are other examples that come to mind. One social historian, Joseph Grusfeld, states that prohibition failed because its supporters did not suffer any actual personal grievance due to others' use of alcohol. Rather, prohibitionists were more motivated by a desire to affirm their own cultural values than to avoid harms to themselves. This provides a cautionary story for climate activist organizations that are largely lacking climate victims among their members.

The civil rights movement also employed boycotts and abstention, most famously in the Montgomery bus boycott. African American victims of racial injustice boycotted the bus system that enforced segregation. These boycotters walked or car-pooled to work for an entire year in order to inflict economic and political pressure on an unjust institution. This action was both individual and collective: individuals changed their behavior to opt out of an unjust transportation system, but they did so as part of an organized effort that took advantage of existing community-based church organizations. The sense of shared sacrifice fortified individual commitment and multiplied the political and economic forces that were brought to bear on the privately operated bus system. Church and activist organizations supported the boycotters morally by publicly praising the "walkers," and practically by organizing car pools and volunteer drivers for those who could not walk.

Citizens concerned about climate in the twenty-first century could also put economic and political pressure on an unjust fossil fuel industry by walking and carpooling to work to avoid being its customers. Such individual efforts directly lead to increased well-being for anyone who opts out of the fossil fuel economy. But individual action becomes a force multiplier when it is shared through social and activist networks and becomes collective action. Effective individual action requires participation in collective efforts.

Ultimately, abolition was successful in transforming an unjust economic system with a long history, and the civil rights movement was successful in ending legally enforced segregation. But the effort to achieve full equality for African Americans, like prohibition, is also a cautionary tale. The Thirteenth Amendment, abolishing slavery, was only adopted after a bloody civil war. And the promise of abolition took a century to accomplish. Cultural resistance in the South in the form of peonage, Jim Crow laws, and prison labor kept the vast majority of African Americans in the former slave states in a condition of servitude until the middle of the twentieth century, and some of these practices, such as prison labor and the mass incarceration of African American males, persist to this day.

Like the Thirteenth Amendment guarantee of freedom from servitude, the Fourteenth Amendment guarantee of racial equality and the Fifteenth Amendment guarantee of voting rights took a century to come to fruition and are still a work in progress.

Government policy is a top-down catalyst to change. Laws are enacted at the national level that direct people and businesses to change the way they manage their affairs, with whom they can (or must) transact business, and what products or services they may buy or sell. In the case of abolition, the Thirteenth Amendment abolished an entire category of property (property rights in human beings). That category of property, now universally recognized to be morally repugnant, had previously been held by the Supreme Court to be constitutionally protected.

But top-down regulation, no matter how well informed and designed, may be limited in its effectiveness if it is not accompanied by underlying bottom-up culture change. Prohibition was ineffective because it was largely flouted and the political culture lacked the commitment necessary to enforce it. The abolition and equal protection amendments took a century to implement because deeply embedded economic structures and cultural attitudes in the former Confederate states did not accept the underlying principle of equal rights.

The 1954 *Brown v. Board of Education* school desegregation decision provides another example of a failure of government policy to accomplish cultural change. While widely viewed as a turning point in racial justice in the United States, its promise is largely unfulfilled sixty-five years later. Public education remains largely segregated by race: in 2011, only 23 percent of African American public school students attended majority-white schools. While *Brown v. Board of Education* established a constitutional policy of integration, American culture has not, even now, adopted integration in fact. Even a city as diverse as New York City has public schools that are among the most segregated in the nation.

As the failed example of prohibition and the incomplete success of the civil rights reforms illustrate, the relationship between legal reform and the underlying social movement is key to the success of remedial laws. Richard Kluger, a critical historian of the desegregation effort, identified the challenge of changing culture through policy:

> But law, in a democracy, cannot impose that resolution by the force of the state alone. Democracy is too unruly for that. That is its great weapon against the would-be tyrant; that is the agony it imposes on the most enlightened reformer. Law in a democracy must contend with reality. It has to persuade. It has to induce compliance by its appeal to shared human values and social goals. How well law succeeds in winning, however reluctantly, the abandonment of unjust private advantage is perhaps the severest, and best, measure of that society's humanity.

INDIVIDUAL CLIMATE ACTION AS PERSUASION

To be successful, then, policy must not just prescribe change; it must persuade people that change is worthwhile. For policy to be effective, culture must become receptive to the change. Individual consumer actions can play a role in persuading people that change is needed, bringing about the necessary culture change to make climate policy effective.

There are some relatively recent examples of how government policy succeeded only after consumer choice laid the groundwork. Bans on smoking in public establishments and the phaseout of chlorinated fluorocarbons (CFCs) were accomplished by policy change but were facilitated by consumer choice.

Smoking used to be ubiquitous in American life. The "smoke-filled bar" was a cultural cliché. But public awareness of the dangers of tobacco smoking led to a decline in smoking in the latter part of the twentieth century, despite the considerable efforts of the tobacco industry to cast doubt on the link between smoking and cancer. Smoking inside and in public places stopped being cool, and consumers began to demand nonsmoking areas in restaurants, bars, and public transportation. Public health agencies identified secondhand smoke as a significant health risk for nonsmokers, leading to bans on smoking in aircraft and other public transportation facilities. Despite dire predictions of the end of the restaurant and bar industries, smoking bans have been widely accepted and implemented by states and municipalities across the country.

Similarly, the well-publicized discovery of the ozone hole over Antarctica led to public awareness of the role that chlorinated fluorocarbons were playing in altering the global atmosphere. These fluorocarbons were widely used in consumer spray products such as hairsprays and deodorants. Long before these substances were banned, consumers began to demand CFC-free products. This consumer acceptance helped lead to the smooth implementation of the international phaseout of CFCs under an international agreement known as the Montreal Protocol.

Smoking bans were effective because smoking became socially unacceptable in large parts of American culture. The phaseout of CFCs was effective because consumers voted against products that contained CFCs by purchasing alternatives. It helped that smoking was not considered a necessity of life and that alternatives to CFCs for propellants and refrigerants were available. But, as this book details, alternatives to fossil fuels are also available for all of the necessities of life. Limits on fossil fuels will be much more likely to be adopted and effectively implemented when using fossil fuels becomes uncool and large numbers of individuals are demonstrating that you can live without them by choosing nonfossil alternatives.

Individual climate action thus has a role to play in convincing society at large that change is appropriate and workable. Social psychologist Per Espen Stoknes, in his book *What We Think About When We Try Not to Think About Global Warming*, argues that because of peoples' natural psychological defenses, scientific facts and figures and predictions of doom are not likely to change peoples' minds about the need for climate action. Rather, positive stories about individual actions in the face of climate change are more convincing.

If you are reading this book, chances are that you are more motivated than most of the people around you to take action to fight climate disruption. That makes you a climate leader. Whether you think of yourself as a leader or not, you are ahead of your social group. Taking individual action to live consistently with your beliefs gives you a positive story to share with your family, friends, and coworkers. Your friends look around them for social cues about which consumption patterns are admirable and which ones are not. You can be part of the movement to establish new cultural norms that will allow for effective climate policy.

LEGISLATIVE APPROACHES TO REDUCING CARBON EMISSIONS

There are two basic legal approaches to reducing harmful pollution like carbon emissions—regulation and economic incentives.

Direct regulation is what usually comes to mind when we think of environmental regulation. The Environmental Protection Agency (EPA), with authority from Congress, sets permissible rates of pollution for individual facilities. This approach works well when there are affordable, available technologies for reducing pollution emissions, and EPA can require all plants in a given industry, or all products in a given category, to install this technology or its equivalent. This technology-based approach has been very successful in reducing water pollution from sewage using secondary treatment and air pollution from cars using catalytic converters.

But this direct regulation approach has been less successful at meeting hard ecological limits on pollution emissions. Both the Clean Air Act and the Clean Water Act set deadlines for EPA and the states to regulate air and water pollution to ensure clean air and swimmable and fishable water, and these deadlines passed decades ago without being met. The problem is allocating the necessary reductions because doing so involves adding more costs for some people and industries than for others—a politically thankless task that smacks of a centrally planned economy. State and federal regulators preferred to let the deadlines pass than to take the political heat for choosing who must reduce their pollution.

Economic incentives, on the other hand, enlist the marketplace to allocate needed reductions. Carbon taxes and cap-and-trade schemes are both economic incentives. In order to emit a ton of a given pollutant—say, CO_2e—you either have to pay the tax or pay to buy an emissions allowance. Money collected from either approach can be applied to general revenues or returned to the people or industries most hard hit by the pricing scheme, as policymakers see fit; in the case of cap and trade, the allowances can be distributed for free.

A cap-and-trade system ensures that the limit used to set the cap will be met, while the market price as allowances become scarce is less certain. A carbon tax provides certainty about the price for emissions, while the amount of pollution reduction actually achieved at that price will be determined by the marketplace reaction to the price increase. Because of the price certainty, economists prefer carbon taxes. Because of the certainty of meeting ecological limits, environmentalists should prefer a cap-and-trade system. In either case, the ultimate market price per ton of emissions needed to achieve a given overall limit should, at least in theory, be the same. Estimates of the carbon price needed to achieve an 80 percent reduction in emissions are as high as $1,000 per ton—the equivalent of $10 extra per gallon of gasoline.

Subsidies for nonfossil energy also work as an economic incentive—as sort of an inverse carbon tax. Instead of making fossil fuels more expensive to the consumer, they make nonfossil fuels seem cheaper. Since the subsidy is hidden from the consumer, it does not appear to cost anything. But someone has to pay for the subsidy, either by increased government borrowing, taxes, or reduced spending on other programs.

If economic incentives do not work in time, there is one form of direct regulation that avoids the allocation problem. That is a direct ban, like the ban on hunting endangered species. An equivalent ban would be the abolition of fossil fuels.

REFERENCES

Attari, Shahzeen Z., David H. Krantz, and Elke U. Weber. "Statements About Climate Researchers' Carbon Footprints Affect Their Credibility and the Impact of Their Advice." *Climatic Change* 138, nos. 1–2 (2016): 325–338.

Blackmon, Douglas A. *Slavery by Another Name: The Re-enslavement of African Americans from the Civil War to World War II*. New York: Anchor Reprint Edition, 2009.

Branch, Taylor. *Parting the Waters: America in the King Years, 1954–1963*. New York: Touchstone, 1988.

Childress, Sarah. *A Return to School Segregation in America?* PBS, July 2, 2014. http://www.pbs.org/wgbh/frontline/article/a-return-to-school-segregation-in-america/.

Clark, Norman H. *Deliver Us from Evil: An Interpretation of American Prohibition*. New York: Norton, 1976, 4–5.

Drescher, Seymour. *Abolition: A History of Slavery and Antislavery*. New York: Cambridge University Press, 2009, 80–81.

Faulkner, Carol. "The Root of the Evil: Free Produce and Radical Antislavery." *Journal of the Early Republic* 27 (2007): 377–405.

Garrison, William Lloyd. "Products of Slave Labor." *Liberator*, March 5, 1847.

Garrison, William Lloyd, and Isaac Knapp. "Peirce's Free Grocery Store." *Liberator*, June 11, 1831, 95.

Gause, Lea W. "The Products of Slave Labor." *Liberator*, April 9, 1847, 59.

Gold, Mike. "Climate Change Super Villains." *Riverdale Press*, December 31, 2014. http://riverdalepress.com/stories/Climate-change-super-villains,56073.

Gusfield, Joseph. *Symbolic Crusade: Status Politics and the American Temperance Movement.* 2nd ed. Champaign, IL: Illini, 1986, 7.

Hertsgaard, Mark. "The Making of a Climate Movement." *Nation*, October 22, 2007.

Kluger, Richard. *Simple Justice*. New York: First Vintage, 2004, 745.

McKibben, Bill. "The Case for Fossil-Fuel Divestment." *Rolling Stone*, February 22, 2013.

Oreskes, Naomi, and Erik M. Conway. *Merchants of Doubt*. New York: Bloomsbury, 2010.

People's Climate March. "The People's Climate March Lineup." Accessed September 29, 2018. http://2014.peoplesclimate.org/lineup/.

"The Products of Slave Labor." *The Liberator*, January 4, 1834, 2.

Rosenberg, Gerald N. *The Hollow Hope: Can Courts Bring About Social Change?* 2nd ed. Chicago: University of Chicago Press, 2008, 71.

Stoknes, Per Espen. *What We Think About When We Try Not to Think About Climate Change*. White River Junction, VT: Chelsea Green, 2015.

Svoboda, Michael. "From Social Change to Climate Change: Lessons from the 1960s?" *Yale Climate Connection,* January 9, 2014.

Vermont Anti-Slavery Society. "Second Annual Report of the Vermont Anti-Slavery Society with an Account of the Annual Meeting Holden in Middlebury, February 16 & 17, 1836." http://www.townofmiddlebury.org/document_center/Selectboard%20Meeting%20Packets/2018/15%20-%20June%2012%20SB%20Packet/05%20-%20Antisalvery%20report.pdf.

Passages in this chapter have previously been published in Karl Coplan, "Fossil Fuel Abolition: Legal and Social Issues," *Columbia Journal of Environmental Law* 41, no. 2 (2016): 223–312.

CARBON DIARY

October 2015

October is usually the most spectacular month of the year to be outdoors in the Northeast, when the brief blaze of autumn leaves makes its appearance. This year, October was warm and dry, and I looked forward all month to that one morning when the early sun on the yellow leaves would light my way through the Tarrytown Lakes bike path on my way to work. With the warm weather, we did not have to think about heating the house this month.

The first weekend of October I attended a professional conference at Vermont Law School. Robin joined me for the drive, and we took advantage of the trip to have dinner with our daughter in Amherst on the way up. The conference was timed for the New England foliage season, but I had to spend an entire Saturday indoors at panel sessions. On Sunday, we made a side trip on the way home to hike in the White Mountains of New Hampshire with some of the other conference attendees, taking in the Welch Mountain loop near Plymouth.

We spent the next Saturday night on our sailboat, anchored off of Hook Mountain in the Hudson River, though we had to motor up from Nyack because there was no wind.

On October 17, we spent a lovely fall day in New York City, after I attended the morning sessions of the annual meeting of the American College of Environmental Lawyers (ACOEL). We walked down the length of the High Line Park right to the Whitney Museum, where we took in an exhibit of colorful jazz-era paintings by Archibald Motley. The American College of Environmental Lawyers sponsored an evening jazz band dinner.

The author's Smart car.

Photo courtesy of Karl Coplan.

But a zero-carbon disaster almost happened this month. On Monday, October 19, I started home from my office in the evening, riding my electric motorcycle. At the intersection of North Broadway and the eastbound I-287 exit, the light was green in my direction, so I kept my speed at around 30 mph. I saw a car turning into my path and I thought, wow—I am really going to have a serious accident on my motorcycle—I am certainly going to die—I am hitting the car—wow, I am actually flying over a car—here comes the pavement—hey, I am not dead. I was very lucky; three hours in the emergency room confirmed that I didn't break any bones, just sprained both hands badly. Robin makes me wear the best protective gear when I ride, and it probably saved my life. But my electric motorcycle did not fare nearly as well. After it was towed to the Zero dealer up in Fort Montgomery, the dealer and the insurer confirmed that the bike was a total loss, with bent forks and a cracked frame. While I debated whether to replace the motorcycle, I decided it was time to test-drive a slightly safer form of electric transport.

For years I took pride in the fact that I did not even own a car (though I freely admit to borrowing Robin's hybrid Prius when I needed one). But maybe it was time to look at the current generation of electric cars. A Tesla was financially out of the question. But the Nissan Leaf and the Smart Electric Drive were about the cost of an ordinary fossil fuel–powered sedan. There weren't any Leafs around locally to test-drive, so on the Saturday after my motorcycle accident I rode my bicycle to Englewood, New Jersey, to the Mercedes-Benz/Smart dealership to test-drive a Smart. Robin met me there on her way back from delivering a lecture in New York City. It cracked me up to walk into the palatial Mercedes dealership to look at one of the smallest (and, it turned out, one of the least expensive) cars on the market in the United States. At least they let me park my bicycle behind the reception desk, as their sales manager is an avid cyclist. He was intrigued by the classic Da Rosa bicycle I was riding (a hand-me-up gift from my younger brother).

IS A VEHICLE LEASE EVER CLIMATE RESPONSIBLE?

The Smart electric vehicle (EV) was the first vehicle I ever leased; all the other cars I've had I've owned outright and pretty much driven into the ground. Some people argue that trading in a car and leasing a new one—even an EV—causes more climate harm than good because there are substantial carbon emissions "embedded" in the manufacture of the vehicle. By this accounting, the embedded carbon in your old vehicle is wasted when you trade it in. Likewise, a short-term lease epitomizes our wasteful throwaway culture by assuming you will replace your car every three years. But this is an accounting issue; after all, the car you trade in will be resold and its embedded carbon will be spread over its remaining street life. In financial accounting, the embedded cost of durable goods is "amortized" (spread out) over the useful life of the item. An accounting system for embedded carbon would similarly amortize the carbon cost of a car—so the first three years of its use by me and the last three years of its use by me or someone else would be the same carbon cost whether I kept the car or not. My "direct footprint" approach does not attempt to keep track of this embedded carbon. And in my case, the three-year lease for about $1,500 a year just made a lot more sense than buying the car for $30,000 and driving it for fifteen years. And I didn't have a trade-in.

I wasn't planning to buy a Smart car that very day, but the salesman pointed out that they had a whole inventory of 2015 Electric Smart cars in pretty much any choice of color. They were trying to get rid of them with a really inexpensive lease deal—$129 down and $129 per month for a three-year lease. I was sure this was a come-on deal and that there would be hidden options, like the radio would cost an additional hundred dollars a month. But for the only time in my life, an auto salesman offered to let me drive away in a new car for a price that was less than the bait-and-switch prices offered in the advertisements. It looked like electric cars had come to the point of being as practical as fossil-powered ones. Though it is tiny, the Smart car has full-sized seats, plenty of headroom for my six-foot frame, and more than enough range to get me to work and back on a single charge. And, given my recent experience, sitting in a steel cage (even a small one), surrounded by air bags, had to be safer than riding on two wheels in the open air. I drove a bright yellow and black Smart EV home that day and promptly nicknamed it Bumblebee.

CUT YOUR HOUSEHOLD VEHICLE FOOTPRINT BY 80 PERCENT WITHOUT SPENDING MORE OR RIDING A BIKE TO WORK

Given my near-death motorcycle experience, I am not about to recommend that people trade their sedan for an electric sport bike, though it does get the equivalent of 300 miles per gallon!. And as much as I enjoy my paddle-and-pedal commute (see chapter 1), it's not for everyone.

But there are painless ways to cut your vehicle footprint right now. The typical U.S. household owns two cars, and the average car gets about 24 miles per gallon and gets driven about 13,000 miles per year. This works out to about 11 tons of CO_2 per household.

Want to cut that figure by 80 percent? When the time comes, trade one of those vehicles for a 50-miles-per-gallon hybrid and trade the other one for an electric vehicle. EVs have plenty of range for the typical commute, and you will still have a hybrid for going on those long family road trips and shuttling kids to soccer. With a renewable energy contract, your EV's operating footprint is zero, and your hybrid's annual footprint will be about 2.5 tons—an 80 percent reduction.

You say you can't afford an EV? Look again. While the Tesla Roadster is a rich man's toy, the Chevy Bolt and Nissan Leaf are priced at the mid-market. My Smart-for-Two EV was just about the cheapest lease deal on the market at the time. Unfortunately, the automaker has announced that it is withdrawing the Smart EV from the U.S. market after 2019, but many vehicles in mint condition are available used for well under $10,000.

You say the Smart EV looks funny? Decide what personal lifestyle statement is more important to you—your automotive tastes or your commitment to fighting climate change!

The last Friday of the month, I did get back in my kayak to paddle across the river to work, even with my still-sore hands. The morning sun lit up the golden trees along the way, and my October was complete. The next day, we drove Robin's car to Harrisburg, Pennsylvania, to visit our son Justin and see his new house. We stopped on the way for a hike in the glowing woods of Ramapo Mountain State Park in New Jersey.

My carbon footprint for the month of October was 460 pounds. Big-ticket items this month, in addition to the natural gas for cooking and hot water, were the drive up to northern New England for the conference and hike, several trips by car to New York City for the ACOEL events, being driven to work when I missed the bus the day after my accident, and the drive to Harrisburg. So far, I am still on track for a 4-ton annual carbon budget.

3

SOME CLIMATE BASICS

What We Mean by "Carbon Footprint," How We Measure It, and Why It Matters

This is a book about defining a sustainable individual carbon footprint and living well in modern America within that sustainable footprint. This is not a book on climate science; there are plenty of good books on that subject out there. But some basic climate science is needed to understand the premises of this book. Throughout these chapters I try to convert my daily activities into "pounds" and "tons" of "carbon dioxide equivalents," and I loosely refer to these impacts as my "carbon footprint." These are the terms used by the United Nations Intergovernmental Panel on Climate Change (IPCC) and by environmental organizations and policy experts who speak about climate change. *Carbon footprint* has become a shorthand for the greenhouse gas impacts of human activities. But the term "carbon footprint" is actually misleading: there is no climate harm from carbon itself. Only the oxidized form of carbon—carbon dioxide—is a greenhouse gas. It is also hard for non-scientists to conceptualize impacts measured in terms of pounds or tons of gases in the atmosphere, since most people don't think of gases as having weight. But they do have mass.

WHAT ARE GREENHOUSE GASES, AND HOW DO THEY CAUSE CLIMATE CHANGE?

The earth's atmosphere consists of about 78 percent nitrogen, 21 percent oxygen, and nine-tenths of 1 percent argon. The remaining 0.1 percent of gases are considered "trace" gases, and the lion's share of these consist of carbon dioxide. Carbon dioxide constitutes about 0.04 percent of the gases in the atmosphere and is considered a "trace" gas. In addition to the relatively stable share of these gases in the atmosphere, the atmosphere contains water vapor, which can vary from 1 to 4 percent of the atmosphere.

Scientists have long known that some of the gases in the atmosphere, including carbon dioxide, retain heat. By the middle of the nineteenth century, physicist John Tyndall had proven the heat-retaining properties of gases such as carbon dioxide and water vapor with an apparatus that directly measured the heat-blocking properties of these gases. By the end of the nineteenth century, a Nobel Prize–winning Swedish physicist, Svante Arrhenius, had calculated that the human burning of coal would increase global temperatures (or offset any natural cooling) because of the increased concentrations of carbon dioxide and water vapor.

These gases work like the glass windows in a greenhouse: they allow sunlight in but block the longer-wave infrared radiation that carries heat. So the sunlight comes in and warms the earth, and some of that warmth can't bounce back into space because the heat radiation is blocked by these greenhouse gases. Without these heat-retaining properties, the earth would be too cold to have liquid water and sustain life. The balance between heat reaching the earth in the form of sunlight and being reflected back into space has been relative stable for tens of thousands of years, since the last ice age. But at other times in the earth's history—before human beings had evolved—the earth has been much hotter and much colder. The hotter periods have been linked to higher concentrations of carbon dioxide in the atmosphere, and the colder periods have corresponded to lower concentrations of carbon dioxide in the atmosphere.

It may seem unlikely that a trace gas like carbon dioxide, which is only a fraction of 1 percent of the earth's atmosphere, can have such a big impact on global climate. But the physics of carbon dioxide's capacity for heat retention has been well established for over a century. Think of carbon dioxide as if it were a trace amount of strychnine in your blood: just an 80 parts per million concentration of strychnine in the bloodstream is fatal. Or think of adding a drop of ink to a glass of water: the ink will then block visible light just like trace amounts of carbon dioxide will block heat. The nitrogen and oxygen in the atmosphere are transparent to the heat radiation, but the carbon dioxide, like the ink in a glass of water, is not. And the warming effect of increased carbon dioxide in the atmosphere is magnified by the increased concentration of water vapor that follows. Carbon dioxide is the trigger for change, but water vapor has a multiplier effect because it is a much more potent greenhouse gas than carbon dioxide. Warmer air holds more water vapor. So a small increase in earth's temperature caused by carbon dioxide concentrations allows the atmosphere to hold more water vapor, which causes a bigger temperature increase.

Other greenhouse gases are also important to the earth's heat energy balance. Methane, also known as natural gas, is a potent greenhouse gas that is thirty times more powerful at trapping heat than carbon dioxide. Nitrogen oxides are greenhouses gases, and so are chlorinated fluorocarbons, artificial chemicals that used to be used as refrigerants.

UNDERSTANDING THE CARBON ENERGY CYCLE

So, if carbon dioxide is a greenhouse gas that will increase global temperatures, why do all of the books and websites about reducing your climate impact talk about your "carbon footprint"? Are carbon and carbon dioxide the same thing? Is carbon also a greenhouse gas? It turns out that *carbon footprint* is a misleading term.

Carbon is not actually a gas, much less a greenhouse gas. At sea level, carbon does not become a gas until the temperature reaches thousands of degrees Celsius (or Fahrenheit). There is no gaseous carbon in the atmosphere. Although carbon and carbon dioxide are related in that carbon dioxide is a compound made up of carbon and oxygen atoms, the two materials have completely different chemical, physical, and greenhouse gas properties. But when scientists and environmental advocates sought to increase awareness of people's climate change impacts, the phrase *carbon dioxide footprint* never stuck, while the phrase *carbon footprint* rolled off the tongue and took off. This book uses the term *carbon footprint*, but it is important to remember that carbon, a solid at earthly temperatures and pressures with no global warming effects, and carbon dioxide, a gas at earthly temperatures and pressures with potentially severe global warming effects, are not the same thing.

That said, the relationship between carbon and carbon dioxide is also key to understanding how the challenge of global warming is linked to human energy use. Carbon becomes carbon dioxide when you oxidize it. Oxidization is the chemical process of adding oxygen to an element to make an oxygen compound. So when you add oxygen to carbon, you get carbon dioxide. You also get energy. All life forms in the animal kingdom, including humans, power their bodies by oxidizing carbon in their food with oxygen from the atmosphere and then emitting carbon dioxide. We call it breathing.

With all the creatures on earth converting carbon into carbon dioxide, the planetary ecosystem needs some way to turn that carbon dioxide back into carbon and oxygen. Plants, through photosynthesis, have been performing this trick in perfect balance with the animal kingdom for eons. Plants use the energy from sunlight to suck carbon dioxide out of the air and convert it into carbon in their stems and leaves, and they also produce oxygen, which is cycled back into the atmosphere. The carbon in the plants then becomes food for animals in the food chain, and it keeps getting recycled into carbon dioxide and back again in an endless cycle. That's why breathing doesn't add to global warming: all that carbon you are converting to carbon dioxide had to be converted by plants in the food chain. Until the Industrial Revolution, that cycle was in nearly perfect balance and carbon dioxide concentrations in the atmosphere remained relative constant, reaching about 280 parts per million.

Fire is a form of rapid oxidization. So another way to convert carbon into carbon dioxide is to burn the stuff. There is carbon in wood, peat, coal, oil, and natural gas. When we burn these materials, we generate heat. The Industrial Revolution came about when John Watt perfected steam engines to convert heat energy into mechanical energy to run mills and railroads. The race was on to burn all the coal we could dig out of the ground. The internal combustion engine allowed us to burn petroleum in more compact vehicles, allowing the development of the automobile. Either fuel could be used to run generators to generate electricity. These fuels powered the great global industrialization.

But all that energy has to come from somewhere. One of the fundamental laws of science—the First Law of Thermodynamics—is that energy can neither be created nor destroyed in a closed system, just converted from one form to another. You can think of elemental carbon, and the carbon bound up in hydrocarbons like oil and gas, as stored energy. The energy in these carbon-based fuels originally came from the sun and was stored by plants through photosynthesis. Unfortunately for the world's energy and carbon cycle balance, the solar energy stored in fossil fuel reserves took hundreds of millions of years to accumulate.

When you burn these carbon-based fuels, the stored solar energy is released in the form of heat and mechanical energy. Carbon dioxide gas—the prime global warming culprit—is the natural byproduct of burning all that carbon and harnessing the stored energy. Carbon dioxide can be turned back into atmospherically harmless carbon by splitting off the oxygen atoms, but that takes exactly as much energy as was released by burning the carbon in the first place—a zero sum game. Photosynthesis can perform this reconversion naturally, but it would take the same millions of years of photosynthesis that went into creating these fuels in the first place.

This energy equation is the reason that addressing global warming from carbon dioxide is such a difficult problem. Carbon dioxide is not some unfortunate byproduct of burning fossil fuels for energy; it is the key chemical product of fossil fuel–derived energy. You just can't take the energy out of fossil fuels without creating carbon dioxide gas, and (for the most part) you can't turn the carbon dioxide gas back into harmless carbon compounds without putting just as much energy from some other source back in.

WHAT IS A TON OF CARBON DIOXIDE, ANYWAY?

So carbon dioxide gas—not elemental carbon—is the global warming culprit, and it is produced by the release of stored solar energy. Other gases emitted by industrial activities contribute to global warming, such as methane, oxides of nitrogen, and fluorocarbons used as refrigerants. But even though these other gases are significant contributors to global warming, and some (like methane) are much more potent greenhouse gases, global climate scientists focus on carbon dioxide gas as the yardstick to measure human-induced climate change because there is so much more of it being emitted than the other greenhouse gases. The term CO_2e is used to describe the global warming potential of other gases in terms of CO_2—the *e* is for "equivalent."

EASY CARBON SAVINGS HINT: ZERO OUT YOUR ELECTRICITY FOOTPRINT

According to the EPA, the average U.S. household uses about 12,000 kilowatt-hours (kWh) of electricity per year, and each of these kilowatt-hours causes the emission of an average of 1.2 pounds of CO_2. So the average conventionally powered house is adding about 7 tons of greenhouse gases annually due to electricity consumption.

You can zero out this big part of your carbon footprint painlessly by switching to a renewable energy supplier that sells wind or solar power. Your local utility still delivers the electricity, but the energy you are buying is carbon free. It might add a few cents per kilowatt-hour to your bill (maybe $10 or $20 per month), but you support the transition to renewable energy and get to count your own electricity footprint as zero.

But wait, you say, aren't the electrons coming to your house still generated at that coal-fired power plant down the street? Not really—it's a scientific fact that in alternating current electrons don't really travel farther than the nearest distribution transformer. What you are buying is pure energy from a vast interconnected grid, and some of the energy on that grid is renewable energy. In many parts of the country you can choose to buy renewable electricity that is on the same grid as your house.

There are two related ways of thinking about how much carbon dioxide is in the atmosphere. We can look at the concentration of carbon dioxide gas in the general atmosphere. Since carbon dioxide is a trace gas, scientists measure this concentration in parts per million instead of percentages. One part per million is one ten-thousandth of 1 percent. Current atmospheric concentrations are about 410 parts per million, or about four-tenths of 1 percent. This is already dangerously high.

But in addition to worrying about the concentration of carbon dioxide in the atmosphere, we have to worry about the rate that human activities are pumping carbon dioxide into the atmosphere and increasing that concentration. So scientists measure greenhouse gas emissions in terms of tons of carbon dioxide. An English ton is equal to 2,000 pounds. A metric ton is equal to 1,000 kilograms, or about 2,200 pounds. Measurements of other greenhouse gases like methane are converted into the equivalent number of tons of carbon dioxide based on how much more potent they are at trapping heat than carbon dioxide.

It may seem odd to measure a gas like carbon dioxide in terms of pounds or tons, measures that are usually associated with weights. Gases seem to be weightless, so how can we measure a "ton" of carbon dioxide? You might remember from high school physics that mass and weight are different but related concepts. Mass is the amount of matter, while weight is the measurable gravitational force exerted by that matter. Gases have mass even though most seem to have no measurable weight when they are floating in the sea of atmospheric gases. Measuring the mass of carbon dioxide emissions is a convenient and precise way of tracking these emissions and calculating their effect on the global climate.

It may seem easier to think of gases in terms of their volume. However, the volume of a given quantity of gas varies wildly depending on the temperature and pressure. So a given amount of carbon dioxide will have different volumes depending on whether you are measuring it at sea level or high up in the mountains, in summer in a warm climate or in winter in a cold climate, or even when a low-pressure weather system is passing through. Volume changes radically, but the mass of a given amount of carbon dioxide is constant.

The IPCC has calculated the amount of carbon dioxide (and carbon dioxide equivalents) that can be emitted from human activities while (most likely) avoiding a global temperature increase of more than 2°C, or about 3.6°F. These amounts are expressed in terms of mass—tons of carbon dioxide. Most carbon calculators likewise use this standard yardstick of global warming potential. So this book follows this convention and refers to greenhouse gas emissions in terms of the equivalent mass of carbon dioxide: tons, or for keeping track of day-to-day activities, pounds.

Still, it might help to think of tons and pounds of carbon dioxide in more accessible terms. One ton of carbon dioxide gas at room temperature and standard sea level pressure would fill a 1,250-square-foot, one-story, 13-foot-high house. That happens to be about the size of the house I live in. Think of a ton as one small houseful of carbon dioxide. Five and a half tons of carbon dioxide—a little more than the annual carbon footprint I am aiming for—would fill an Olympic-size swimming pool. And the average American carbon (dioxide) footprint would fill about three Olympic-size pools. But only at sea level and room temperature.

HOW MUCH MORE CARBON DIOXIDE CAN THE PLANET TOLERATE?

So far the climate and carbon dioxide questions addressed in this chapter have been relatively simple and have clear answers. But the big question—how much carbon dioxide can the planet tolerate?—gets more complicated. The clear and simple answer to that question is qualitative: "nowhere near as much CO_2 as industrialized society is pumping into the atmosphere." But once you try to quantify the answer and set a global or personal carbon budget, the answer gets bogged down with uncertainties, contingencies, and risk.

Part of the problem is that the answer depends on how you define the question; that is, what is a "tolerable" or "acceptable" amount of global warming? If you are looking for zero impact on global temperatures compared to what they would be in the absence of industrial emissions, it's too late already. Atmospheric CO_2 concentrations have reached over 400 parts per million, a concentration they had not reached since over 2 million years ago, long before humans walked the planet. The planet has already warmed by 1°C (1.8°F) over preindustrial times.

The global target for limiting the most catastrophic effects of climate change is an increase of 2°C (3.6°F) over preindustrial times. This limit was established by the United Nations IPCC, and it was adopted as the basis of the emissions reductions contemplated by the 2015 Paris Agreement.

But this 2°C threshold does not avoid severe adverse climate impacts: small island nations will still be swallowed by rising seas, and droughts, crop failures, severe flooding, and famines will still occur. For this reason, the Paris Agreement also adopted a 1.5°C target as something to aim for. Still, limiting warming to 2°C would most likely avoid the most catastrophic, civilization-threatening impacts.

But when you try to convert that 2°C limit to a numerical limitation on tons of CO_2, things get even more uncertain. Climate models can predict with a high degree of certainty that the global climate will warm due to increased CO_2 in the atmosphere and that this warming will be significant. It's also easy to calculate exactly how much each ton of CO_2 emitted into the atmosphere increases the overall CO_2 concentration. The models are not as certain, however, about the exact amount of warming that will result from a given CO_2 concentration in the atmosphere. It is sort of like the models used to predict the weather. Sometimes the best they can do is give the probability that it will rain tomorrow—say, "a 30 percent chance of showers." So estimates of a global "carbon budget" to achieve the 2°C limit on warming are just that—estimates based on a probabilistic assessment of the chances that a given CO_2 concentration will avoid a global temperature change in excess of 2°C.

It may seem odd to base significant policy on impacts that can only be stated as probabilities rather than certainties. But probabilistic risk assessment is an essential and routine tool used in regulations designed to avoid costly impacts. There is a certainty that there will be an impact; probabilistic analysis helps determine how many people will be affected and how severely, even if it cannot tell exactly who will be affected. Think of it as an insurance policy. The insurance industry routinely uses probabilistic risk models to determine the correct premiums for an insurance policy.

The global carbon budget popularized by Bill McKibben's climate advocacy group, 350.org, is a limit of 562 gigatons of CO_2 left as of 2013. A gigaton is a billion tons. This budget is based on an 80 percent chance of avoiding a 2°C increase. But the IPCC uses a different budget, which is quite a bit higher, based on just a 66 percent chance of avoiding a 2°C rise. Based on this riskier budget, the world could afford to emit 987 gigatons of CO_2e as of 2014. (To confuse things even further, the IPCC budget is stated in terms of elemental carbon, which has to be converted to CO_2

equivalents.) The more recent IPCC SR15 supplemental report, issued in October 2018, calculates a global greenhouse gas budget of between 420 and 570 gigatons of CO_2e for a two-thirds chance of meeting a more protective 1.5°C global warming limit.

To sum up, the most recent IPCC estimate leaves a global carbon budget of around 500 gigatons between now and midcentury. This leaves just about 15 gigatons per year between now and the middle of the century. But even that estimate gives the planet only a two in three chance of avoiding the most severe climate impacts—so it would be best to reduce emissions below 15 gigatons per year.

REFERENCES

Heinberg, Richard, and David Fridley. *Our Renewable Future*. Washington, DC: Island Press, 2016.

Intergovernmental Panel on Climate Change. Fifth Assessment Report, Summary for Policy Makers, 2014.

Intergovernmental Panel on Climate Change. Special Report: Global Warming of 1.5°C, Summary for Policy Makers, 2018.

Mann, Michael, and Lee Kump. *Dire Predictions: Understanding Global Warming*. 2nd ed. New York: DK, 2015.

Mann, Michael, and Tom Toles. *The Madhouse Effect: How Climate Change Denial Is Threatening Our Planet, Destroying Our Politics, and Driving Us Crazy*. New York: Columbia University Press, 2016.

Weart, Spencer. *The Discovery of Global Warming*. 2nd ed. Cambridge, MA: Harvard University Press, 2008.

CARBON DIARY

November 2015

November is the transitional month in the Northeast, as fall fades into winter. Mild weather continued, and the bright autumn leaves clung to the trees unusually long this year. In November we rake up the leaves, clean the chimney flue, and start cutting up the downed trees in our backyard lot to season them for firewood. This year I discovered several old, dried-out red logs that looked like weathered driftwood even though we are nowhere near water. They burned well. I eventually figured out that they were the hardwood cores of wild cherry trees, the sapwood around the core having rotted away. I gathered up a good windfall of these logs for eventual use heating the house. Over the November weekends, I set about cutting the piles of already seasoned logs to stove length and started splitting them by hand. They say that "wood heat heats twice"—once when you work up a sweat cutting and splitting it, and a second time when you burn it. I just know I like an excuse to be outside on a crisp fall day working with my hands in my backyard, with Clausland Mountain coming into view as the trees lose their leaves.

We spent some time on Sunday, November 8, preparing our boat for the winter, and we stayed on board overnight. During the cold half of the year, when I can't kayak across the river and bike to work, a few times a month I try to bike to the ferry in Haverstraw and on from the ferry landing in Ossining to my office in White Plains. It's a nice way to get exercise and start the day.

Since our boat is at her winter berth in Haverstraw, it was a short ride to the ferry landing on the morning of the ninth. The mists covering Hook Mountain in the morning made a shimmering blanket of fog. I always

Hook Mountain in the morning mist.

Photo courtesy of Karl Coplan.

figured that since my bike-and-ferry commute included about 20 miles of zero-carbon bicycling and just 5 miles on the diesel ferry, the carbon footprint of the ferry portion of the trip would be minimal. But on this particular Monday morning (knowing I was keeping a daily log of my carbon footprint), I decided to look up the carbon impacts of the commuter ferry. I was shocked to find that the New York Waterways ferries had a fuel efficiency of 9.7 passenger miles per gallon—so my 5-mile trip was worth about 10 pounds of carbon. This is not much less than a round-trip drive by myself all the way from my home to my office and back in a hybrid Prius. So biking and taking the ferry doesn't really save any carbon impacts (or dollars) compared to just driving there. Given that the bike-and-ferry trip takes two hours, that's one quirky commute I am likely to give up.

I replaced my totaled electric motorcycle in November; it was just too much fossil-free fun to give up. Zero had a deal on leftover 2014 S model electric bikes, so with a generous trade-in for my totaled wreck and what was left of the insurance payment, I ended up with a newer, faster, farther-ranging, shinier, and better model. So I have a matching fleet of electric yellow electric vehicles. I promised Robin I would be more careful at intersections from now on.

We made one weekend trip up to our cabin in the Adirondacks to get the cabin ready for winter as well. It was a great chance to climb Moxham Mountain in an early dusting of snow, in addition to stacking wood and emptying the composting toilet. We fired up the woodstove at the cabin and made things way too hot inside. By the end of the month, we were firing up the woodstove at our house as well, as the week before Thanksgiving was chilly. No one complained of a cold house on Thanksgiving Day!

My climate footprint for the month was about 290 pounds, much less than October despite the fading hours of sunlight and dipping temperatures. As always, the natural gas for the stove and water heater was the big-ticket carbon item for the month, while our carbon footprint for driving dropped since we only made one weekend trip up north in addition to driving back from Pennsylvania on the first.

4

SUSTAINABILITY

What Is It, and Who Can Really Claim to Be Doing It?

Sustainable is cool, chill, hip, righteous. It is socially responsible. Sustainable is the new black. There is hardly anyone on the planet that does not want to be sustainable. It is the epitome of everything that is desirable among greens and the antithesis of everything that is undesirable. All sorts of consumer products are marketed as sustainable. You can even buy sustainable *disposable* diapers, cutlery, and dinner plates, adding their sustainable mass to the mountains of sustainable landfills around the country. You can invest in sustainable investments. Real estate developers even call their projects "sustainable," as if there were an inexhaustible supply of land to be converted into condominium projects. The magazine *Ad Age* listed *sustainability* in its 2010 list of the top ten "Jargoniest Jargon We've Heard All Year," calling it "a squishy, feel-good catchall for doing the right thing."

We all want to live sustainably. Ultimately, we all *have* to live sustainably, or there won't be any "us" on the planet anymore. But what does sustainability actually mean? Isn't *sustainable development* a contradiction in terms, as the very idea of development implies the consumption of nonrenewable resources like land, minerals, and fossil fuels?

A SHORT HISTORY OF THE IDEA OF SUSTAINABILITY

Since the term *sustainability* seems to mean all things green to all people (and to all product marketers), it is worth taking a look at its history to try to get at the true meaning of the term.

While the environmental movement gets credit for adopting the concept of sustainability in the 1970s, its use as a term for environmental resource management dates back much farther than that, back to the days of Gifford Pinchot and the idea of sustainable forestry. Believe it or not, one of our nation's founders, Thomas Jefferson, invoked the idea of sustainability in his plan for the newly formed United States. Writing to James Madison on September 6, 1789, he set out as a basic principle the idea that the living generation has no right to limit the lives of future generations:

> I set out on this ground, which I suppose to be self-evident, that the earth belongs in usufruct to the living. . . . For if [a member of the present generation] could, he might, during his own life, eat up the usufruct of the lands for several generations to come, and then the lands would belong to the dead, and not to the living, which would be the reverse of our principle.

Jefferson relies on the term *usufruct*, which is legal Latin, an estate law term that refers to the product of lands rather than the lands themselves. In financial terms, *usufruct* is analogous to the interest that a sum of capital might earn, rather than the capital itself. Jefferson is saying that the present generation must live off the ongoing produce of the earth—the "interest" it generates—and not invade and steal its capital. This is the essence of sustainability.

The idea of sustainability, as well as the acknowledgment that the earth's productive capacity has limits, took form a century after Jefferson. In the late nineteenth century, Gifford Pinchot, head of the U.S. Forest Service, pressed for more active forest management: that is, deliberate replanting and harvesting at a rate equal to forest regrowth.

While modern-day environmentalists (and Pinchot's nineteenth-century rival, Sierra Club founder John Muir) scoff at Pinchot's utilitarian and profit-maximizing approach to forest management, his idea of harvesting natural resources at a rate that does not exceed their natural regeneration remains at the core of the concept of sustainability. Indeed, Pinchot's idea of conservation—meaning a form of resource management that could be continued indefinitely—was adopted into national law in the

Taylor Grazing Act of 1934. By 1960, Congress had adopted the concept of sustainability in the Multiple Use Sustained Yield Act, which defined *sustained yield* as the "achievement and maintenance in perpetuity of a high-level annual or regular periodic output of the various renewable resources of the national forests without impairment of the productivity of the land."

By the 1970s, the environmental movement was in full force as a political and social movement in the United States. Earth Day 1970 celebrated—and demanded—recognition of the earth's limits and incorporated the idea, if not the term, of sustainability. Similarly, in 1972, the United Nations Conference on the Human Environment declared, in its Stockholm Declaration, the fundamental principle of preserving environmental values and resources for future generations:

> PRINCIPLE 2
>
> The natural resources of the earth, including the air, water, land, flora and fauna and especially representative samples of natural ecosystems, must be safeguarded for the benefit of present and future generations through careful planning or management, as appropriate.

The consensus that emerged from the Stockholm conference led to the formation of the United Nations Environmental Programme in December 1972.

Ten years later, as people became more aware that global development and poverty eradication needed to be kept in harmony with environmental values, the UN general secretary called upon former Norwegian prime minister and public health professional Dr. Gro Harlem Brundtland to lead a commission on the environment and development. In 1987, the Brundtland Commission issued its report, entitled *Our Common Future*. This report coined the term *sustainable development* and gave it the definition that prevails to this day:

> Sustainable development is development that meets the needs of the present without compromising the ability of future generations to meet their own needs.

The Brundtland Report expanded on this concept:

> Many of us live beyond the world's ecological means, for instance in our patterns of energy use. . . . At a minimum, sustainable development must not endanger the natural systems that support life on Earth: the atmosphere, the waters, the soils, and the living beings. . . .
>
> In essence, sustainable development is a process of change in which the exploitation of resources, the direction of investments, the orientation of technological development, and institutional change are all in harmony and enhance both current and future potential to meet human needs and aspirations.

But the report also acknowledged the parallel goal and human need to eradicate global poverty, asserting, "A world in which poverty and inequity are endemic will always be prone to ecological and other crises. . . . Sustainable development requires that societies meet human needs both by increasing productive potential and by ensuring equitable opportunities for all."

The Brundtland Commission's idea of sustainable development formed the basis for the 1992 Rio Conference on the Environment and Development, and the Rio Declaration adopted by 172 nations incorporated the principle of sustainable development as its guiding goal.

BUT WHAT IS SUSTAINABILITY?

The Brundtland definition of sustainable development masks some serious ambiguities. The simplest understanding of the phrase "meet the needs of the present generation without compromising the ability of future generations to meet their own needs" would seem to require that resources not be used up faster than the natural systems of the earth can replenish them. Under this understanding of the term, fossil fuel use is by definition unsustainable, as the natural systems of the earth do not replenish fossil fuels at the rate that modern civilization burns them. At current rates, humanity is on course to burn in just three centuries all the fossil fuel reserves that the earth took millions of years to generate.

Driving a gasoline-powered car, flying in a jet plane, riding a diesel train, and heating your house with natural gas or oil are all unsustainable in this understanding. So is switching on the electric lights with electricity generated by a coal-, oil-, or gas-fired power plant.

If the current generation does these things to "meet our needs," then future generations will not be able to do these things to "meet their needs." Fossil fuel use is then simply unsustainable, without even considering the fact that the greenhouse gases resulting from our current rate of fossil fuel use will make large parts of the earth uninhabitable and unsuitable for agricultural production because of climate change, keeping those future generations from meeting their basic needs for food and health even apart from their need to drive cars, fly planes, ride in trains, and heat their houses like we do.

The strictest sense of sustainability would thus preclude any use of fossil fuels and compel exclusive reliance on renewable energy. After all, by definition, renewable energy is the only energy that can be sustained indefinitely, since it can be renewed at the same rate it is used. The Brundtland Report, however, did expand the notion of sustainability by including the phrases "the orientation of technological development, and institutional change"; in other words, it suggested that sustainability takes into account the pace of technological change.

This addition suggests that fossil fuel use might still be sustainable so long as it satisfies two criteria. First, fossil fuels must not be burned at a rate that will cause irreversible global warming impacts that harm the ability of future generations to meet their basic needs. Second, the sustainable rate of fossil fuel consumption must be calculated based on the time it will take to phase in currently available and reasonably certain technological advances in energy consumption and production that will completely displace fossil fuels.

The hopeful news is that at least one study has demonstrated that existing technologies can get every state in the country to energy independence using renewable, nonfossil energy sources and existing proven technologies. The bad news is that converting to 100 percent nonfossil energy will take some serious adjustments in how we live, where we live, and the basic infrastructure of our economy. These adjustments will take time—decades at least. That is why most groups advocating for a transition to 100 percent nonfossil energy pick a date around midcentury to accomplish this transition. And, not coincidentally, the 2015 Paris Agreement sets midcentury as a goal to achieve zero net carbon emissions from fossil fuels. The more recent Intergovernmental Panel on Climate Change (IPCC) goal of meeting a 1.5°C limit on warming (see chapter 3) suggests an interim goal of reducing emissions by about half by 2030.

The writing is on the wall, then: net zero by midcentury is both a scientific climate imperative and a global goal. But just declaring this goal does not mean that whatever we do in the meantime satisfies the criterion of sustainability. Making a promise that the next generation will live sustainably while we in the present generation continue to consume fossil fuels and generate greenhouse gases at excessive rates does not make our way of life sustainable in any sense of the word. Rather, the only defensibly sustainable rate of greenhouse gas emissions is one that, collectively, avoids a climate change disaster during the transition to a net zero carbon economy.

The next question is: What is a "sustainable" individual rate of carbon emissions?

REFERENCES

Declaration of the United Nations Conference on the Human Environment. Report of the United Nations Conference on the Human Environment, Stockholm, Sweden, June 1972. https://digitallibrary.un.org/record/523249.

Heinberg, Richard, and David Fridley. *Our Renewable Future.* Washington, DC: Island Press, 2016.

Miller, Char. *Gifford Pinchot and the Making of Modern Environmentalism.* Washington, DC: Island Press, 2001.

Report of the World Commission on Environment and Development: Our Common Future. Oslo, Norway, March 20, 1987. http://www.un-documents.net/our-common-future.pdf.

CARBON DIARY

December 2015

This December was the mildest ever in the New York metro area. The temperature on Christmas Eve topped out at 70°F. We were picking fresh greens from our vegetable garden for our holiday dinner. We have a family tradition of ice skating outdoors on the afternoon before Christmas, but the skating rink at Bear Mountain State Park was closed in the afternoon because of "extreme temperatures," so we all went for a walk in our swamp instead. I didn't even consider turning the heat on in the house until after Christmas Day, though we had the obligatory Yule fire, mostly for decoration.

I had to resolve some carbon-tracking transportation issues during the month. On December 1, I shared a limo into New York City with my Pace colleague and litigation clinic partner, Bobby Kennedy. We were both headed for the Riverkeeper holiday party. He was taking the limo in anyway and then leaving from the party directly for the airport. How do you account for the carbon footprint of a shared ride like that? The easy way out is to say, hey, the limo is going there anyway, so my share is zero, but by that way of thinking, all air travel is carbon free because, hey, that jet plane was going to fly to Los Angeles anyway. So I counted my share of the one-way ride in a midsized limo. I took the train back to my electric car after the party.

I spent one weekend managing my woodpile. Most of the logs I had been seasoning since 2013 were too long for the woodstove insert I installed in 2014, so I spent a day cutting the logs down to size with an electric chainsaw and then moving the seasoned stack of wood onto the porch and close to the house for the winter. I finished splitting the logs to season

The woodstove at the author's ski cabin.

Photo courtesy of Karl Coplan.

for the 2017 winter and then spent some time stacking this wood. I think I can count three or four rounds of wood warmth—cutting, dragging, splitting, stacking, then stacking again—then finally burning it when the cold weather arrives. To some people this all might seem like a chore, but I enjoy the excuse to be doing something outside with my hands in the fall weather, enjoying the occasional hawk flying overhead or the neighbor walking by.

CUTTING YOUR HEATING BILL

Not everyone can cut his or her heating bill by feeding a woodstove with deadwood from the backyard. And it's a good thing, too, since even modern woodstoves emit enough noncarbon air pollutants that populated areas would have unbreathable air if everyone relied on them. Not to mention that there simply aren't enough trees to go around.

But you can cut your home heating carbon footprint with some simple steps. About half the homes in the United States are heated with natural gas. Here in the Northeast, the typical carbon footprint of a gas-heated home is over 10 tons—just not sustainable on a 4-ton-per-person carbon budget. Besides the obvious step of setting your thermostat lower and bundling up, you might consider shifting some of your residential heating load to your carbon-free renewable electricity contract with a plug-in radiator in the rooms that are actually occupied, while keeping unoccupied rooms cold.

In the long run, electric heat pumps running on renewables may be the most carbon-efficient way to heat our homes. If you are already one of the 30 percent of U.S. households that are on electric heat and you have signed up for a renewable energy contract, then congratulations, you have already achieved zero carbon home heating.

On December 11, I tested the limits of my electric motorcycle; Robin left the Prius at her office while she was traveling to California, and I needed to get to Bridgeport, Connecticut, to visit a longtime friend in the hospital, 50 miles from our house. Terry Backer was a founder and hero of the Waterkeeper Movement, a worldwide organization of grassroots advocates for the health of local water bodies. I figured I could stop and charge the bike at my office on the way back. I made it to Bridgeport just fine, and the bike showed just enough range to make it back to my office in White Plains to plug in. But I didn't realize that as the remaining charge on the bike dropped below 10 percent, the controller boards started dropping the top speed. I found myself limping along in "snail mode" at 20 miles per hour on the expressway, in the dark, just a month after having had a serious motorcycle accident back in October. I had to ride on the shoulder to get to White Plains and an outlet for charging.

Terry passed away that next week. The wake and funeral in Norwalk were within range of my electric car and motorcycle, with a little boost charge at the office.

Robin got back from California on December 18, and I went to pick her up (with her Prius) at Newark Airport as a treat. How should I count the carbon impacts of driving to Newark? If I had not picked her up, Robin would have called a limo service and would have made the same trip in a larger car, but that would have been part of Robin's carbon footprint, not mine. (I am not quite ready to take responsibility for my wife's carbon footprint, or my adult children's. I count these as impacts that are not under my control.) I counted the carbon impacts of the round trip anyway; if I get good-spouse credit for driving to the airport to pick Robin up, I suppose that's on my carbon tab, not hers. We drove the next morning up to our cabin in the Adirondacks to sneak in a weekend of cross-country skiing on the thin early-season snow, and my share of that trip is on my carbon tab for the month, too.

We also drove up to our Adirondack cabin the week after Christmas, but my daughter, Beryl, home from college, wanted to stay home for New Year's with her friends. Beryl is not into feeding the woodstove constantly, so I turned the heat on for her, and, in the spirit of the season, put that on my carbon tab, too.

5

WHAT IS INDIVIDUAL CARBON SUSTAINABILITY, THEN?

If we accept that, somehow, the world has to get on track to a zero net greenhouse gas energy economy by midcentury and that, in the meantime, we can't all collectively burn more than the equivalent of about 500 gigatons of carbon dioxide–producing fossil fuels, or about 15 gigatons per year, then what would be a sustainable individual carbon footprint during this time of transition to a zero carbon energy economy? An equal global allocation of this 15 gigatons would be about 2 tons per person per year, or about 1 ton of direct carbon impacts, which is pretty much inconsistent with anything but the most Spartan life in the developed world. ("Direct" impacts are under our control; "indirect" impacts are emissions not under our individual control, such as energy use at our place of employment, emissions in the production of goods we buy, and emissions caused by government activities.) But what if that carbon allocation, like other global goods, were distributed more proportionately to global incomes rather than per capita? That may be inconsistent with the most idealistic goals for global equality, and it may also be inconsistent with the purest forms of distributional justice, but it may be the most realistic allocation.

Since there is no global allocation system for carbon emissions, any attempt to come up with a sustainable individual share of the global limit is going to be hypothetical. There are several ways to think about the problem. One way is to imagine how much of an individual carbon footprint one would be willing and able to pay for if there were actually a price on carbon emissions. Another way would be to consider where one falls in the global distribution of incomes and assume that one might justify a share of sustainable greenhouse gas emissions proportionate to one's share of global income. A third way might be to consider the proportionate reduction necessary for the United States's collective emissions of greenhouse gases and what that would mean for one's individual share. All of these approaches tend to support an individual share of direct greenhouse gas emissions of about 4 or 5 tons annually.

WHAT CARBON FOOTPRINT COULD I AFFORD IF I HAD TO PAY FOR IT?

One way to achieve this proportional allocation would be to increase fossil fuel prices to the level necessary to reduce greenhouse gas emissions by half. This might be accomplished either through a tax on carbon or through a global cap-and-trade system for carbon emissions (see chapter 2). Since carbon-pricing schemes allow wealthier individuals and nations to buy the rights to above-average emissions, a cap-and-trade system or a tax would allocate greenhouse gas emissions roughly according to global wealth, just as most other scarce goods are distributed globally in our market economy. Under a cap-and-trade system, total emissions of an air pollutant (like carbon dioxide and other greenhouse gases) are "capped" at the level necessary to protect the environment. In the case of greenhouse gases, that cap would have to be at the rate that averaged about 15 gigatons annually until midcentury and zero thereafter. That is, to be considered sustainable, a cap would have to assume both a tolerable current rate of emissions and the use of technology to achieve a permanently sustainable level of emissions. With a 15-gigaton annual cap, persons, governments, and any industries engaged in carbon dioxide–generating activities would have to purchase sufficient allowances to cover their emissions, bidding the price of this limited resource up to whatever price it takes to clear the market demand. Since this cap represents a bit less than half of the 35.7 gigatons emitted globally in the year 2014, presumably the price of emissions credits would have to be fairly substantial to convince people to forego greenhouse gas–producing activities and reduce global emissions by more than 50 percent.

An alternative way to achieve equivalent market-based reductions would be to impose an emissions tax instead of a cap-and-trade system. Either way, the system works by raising the price of greenhouse gas emissions until those unable or unwilling to pay for such emissions at current levels reduce their emissions. Estimates of the carbon price needed to achieve a 50 percent reduction in greenhouse gas emissions by midcentury vary from \$200 to \$400 per ton of CO_2e. At \$400 per ton, this equates to adding \$4 to the price of a gallon of gasoline, \$6 a pound to the price of beef, and about \$600 to the cost of a round-trip transcontinental airline ticket. But a 50 percent reduction by 2050 is not nearly enough. A 2014 Stanford University study on U.S. technology and climate policy strategies indicates that, to achieve an 80 percent reduction in emissions, a carbon tax in the year 2050 would have to be between \$100 and \$900 per ton—corresponding to a gas price increase anywhere between \$1 and \$9 per gallon.

So you might ask yourself, "How much less would I drive if I actually had to pay \$7 for a gallon of gasoline?" and "How much less would I fly if the cost of an airline ticket roughly doubled?" You would ask these questions to come up with what you would actually be willing (and able) to pay if greenhouse gas emissions covered a cost commensurate with ecologic limits on their availability. For the average American with a 20-ton annual carbon footprint, maintaining this footprint might cost \$8,000 a year (assuming a carbon price of \$400 per ton), either from direct carbon charges or from increased prices for consumer goods and taxes. Cutting this footprint in half would reduce the cost to about \$4,000. For a family of three (the average U.S. household size), you might multiply these numbers by 3: so, to maintain a family of three, you would have to purchase \$24,000 in carbon emissions rights. You might say, that's outrageous; very few families can afford to pay \$24,000 to maintain their carbon emissions. But that is exactly the point: the only way a carbon tax is going to get people to reduce their emissions drastically is to make a high-emissions lifestyle unaffordable for most people. Most people will have to find other ways to live their lives with less carbon.

So you could set your own personal footprint goal based on what you think you would be willing (and able) to pay if an effective global market for carbon emissions were implemented tomorrow. But this would be a rhetorical question, since you wouldn't actually have to pay the money. (Economists call this sort of inquiry "contingent valuation"—the question of what the public would be willing to pay for environmental resources if there actually were a market for such values.) A more realistic approach might be to donate the money that represents the true price of greenhouse gas emissions to a charity devoted to relieving the suffering of climate victims in low-lying impoverished regions of the world.

How about buying offsets with the money? For reasons I will go into later, offsets don't represent true mitigation of the greenhouse gas impacts of our activities, and they will be vastly underpriced as long as there is no corresponding limit or price on carbon emissions. But anything that represents parting with real dollars would present a more realistic wealth-based limit on your carbon emissions.

And this question is doubly rhetorical, since no global (or national) carbon emissions trading system is likely to be implemented, much less one with a 15-gigaton cap. The United Nations Framework Convention for Climate Change attempted to implement a cap-and-trade system through the Kyoto Protocol, which was adopted in 1997 and went into force in 2005 (and expired in 2012). But the initial goals of the Kyoto Protocol fell far short of the 50 percent reduction in greenhouse gas emissions that might be considered sustainable in the short term, and they called for no reductions by some of the largest global emitters in the developing world. Moreover, the United States refused to agree with the trading and reduction system of the Kyoto Protocol. The U.S. Congress came close to enacting a cap-and-trade measure for greenhouse gases in 2009, when the American Clean Energy and Security Act, sponsored by Representatives Waxman and Markey, passed the House of Representatives. But the measure was opposed by both the left (which was convinced that tradeable emissions righs would just mean more profits for market traders instead of real climate action) and the right (who called it a job-killing energy tax). The bill never made it to the floor of the Senate. The unsuccessful bill called for fairly modest reductions in U.S. greenhouse gas emissions (only a 17 percent reduction by 2020) and was riddled with exemptions and bypass provisions. The United States's voluntary reduction commitments under the 2015 Paris Accord contemplated a 26 to 28 percent reduction in emissions by 2025, compared to 2005.

So how much would you be able and willing to pay for your carbon footprint? The best guess is that a direct carbon footprint of 4 tons per person would cost the average three-person household about $5,000 per year—not without pain for a middle-class American. The current direct individual footprint of 10 tons (half of the 20-ton average American total footprint), on the other hand, would cost more like $12,000 per three-person household. Probably not affordable. This reminds us that the average current U.S. carbon footprint is just not affordable for the world.

WHERE IS YOUR CARBON FOOTPRINT ON A MAP OF GLOBAL WEALTH?

Trying to imagine our carbon footprint if we had to pay for it is one way to estimate the way carbon emissions might be distributed in global markets just like other scarce goods are distributed. Another way might be to figure out where one's own income falls on a chart of global income distribution and then assume that an individual carbon footprint that is proportional to one's share of global income would be sustainable. One would ask the question "If everyone on the globe at my income level had my carbon footprint, would the combined global carbon emissions stay within the 15 gigatons per year that we can afford to emit while we transition to a zero carbon energy system?"

This is a pretty simple calculation once you get past some controversies about how to compare incomes under different economic systems and different costs of living. Global welfare economists like to use something known as "purchasing power parity" (PPP) to adjust for different costs of living. Using this PPP adjustment, global GDP in 2005 was about $56 trillion in U.S. dollars.

So if you had an annual income of, say, $56,000 (pretty close to the median household income in the United States), you might lay claim to one-billionth of the global income, and also lay claim to one-billionth of the global carbon budget. If the world were on a 15-gigaton carbon budget, that would correspond to a 15-ton household carbon footprint. If you shared that income in a family household of three, then each member of that household would have a 5-ton individual carbon budget—though his or her direct footprint would have to be substantially less.

WHAT IF ALL AMERICANS TIGHTENED THEIR CARBON BELTS THE SAME AMOUNT TO AVOID GLOBAL CATASTROPHE?

A third way to estimate a sustainable individual carbon footprint, assuming that global consumption distribution is not about to change, would be to assume that each nation would have to cut its current carbon emissions by about 50 percent over the next decade to put the globe on a sustainable track of staying below 15 gigatons per year between now and mid-century. In 2016, total U.S. greenhouse gas emissions were approximately 6.5 billion metric tons. A 50 percent reduction would equal 2.75 billion metric tons. Divided among the 320 million people in the United States, that would correlate to an average total per capita carbon footprint of about 9 tons. Since our total footprint includes indirect greenhouse gas emissions (emissions not under our individual control), the average direct carbon footprint of an American would be limited to about half that—or about 4 to 5 metric tons of carbon dioxide equivalent per person per year.

So a current individual carbon footprint of about 4 to 5 tons per year can be considered "sustainable" for a middle-class lifestyle in the developed world. At least one other author thinking about this issue has reached the same conclusion. British author George Marshall argues for a 4-ton limit in his book *Carbon Detox*. So is it possible to live—and live well—on a 4-ton carbon budget? I think so, and the next few chapters of this book will explore how you might make a budget, keep track of your carbon emissions, live within that budget—and have enough left over for life-enriching fun.

REFERENCES

Clarke, Leon E., Allen A. Fawcett, John P. Weyant, James McFarland, Vaibhav Chaturvedi, and Yuyu Zhou. "Technology and U.S. Emissions Reductions Goals: Results of the EMF 24 Modeling Exercise." Special issue, *Energy Journal* 35 (2014): 9–31 and fig. 9.

MacKay, David J. C. *Sustainable Energy—Without the Hot Air*. Cambridge: UIT Cambridge, 2009.

Mann, Michael, and Lee Kump. *Dire Predictions: Understanding Climate Change*. 2nd ed. New York: DK, 2015.

Marshall, George. *Carbon Detox*. London: Octopus, 2007.
National Association of Manufacturers. *Economic Outcomes of a U.S. Carbon Tax*. February 26, 2013. http://www.nam.org/Issues/Tax-and-Budget/Carbon-Tax/2013-Economic-Outcomes-of-a-US-Carbon-Tax-Full-Report.pdf.
Regional Economic Models, Inc. and Synapse Energy Economics, Inc. *The Economic, Climate, Fiscal, Power, and Demographic Impact of a National Fee-and-Dividend Carbon Tax*. Report prepared for Citizens Climate Lobby, June 9, 2014. https://citizensclimatelobby.org/wp-content/uploads/2014/06/REMI-carbon-tax-report-62141.pdf.

CARBON DIARY

January 2016

This January was just a few degrees warmer than average and featured a record snowfall in New York City, so I had a chance to see how much we could rely on the woodstove to cut carbon emissions from our gas furnace, as well as try out my little electric Smart car in the snow. Both worked just fine.

The first week of January each year, the Association of American Law Schools holds its annual conference. This year the conference was in New York, so at least I could attend without adding any air travel to my carbon budget. I did have to figure out the best way to commute from my house in Rockland County to midtown each day. I try to avoid driving in rush hour traffic, and I assumed that public transportation would be the carbon-minimal choice. But to make it to the conference center in time for the environmental law field trip on the first day of the conference meant taking a 7:00 A.M. bus from West Nyack across the river to catch a 7:30 train from Tarrytown. The conference field trip involved a subway ride to lower Manhattan and a ferry ride to Governors Island, where we toured New York City's Harbor School, a public high school specializing in marine sciences and marine trades, and heard about New York's plans for rising sea levels caused by climate change, as well as projects to restore oyster reefs to New York harbor. When I totaled up the carbon emissions for the day—all from travel on public transit—I was surprised to find that the bus, train, subway, and ferry added up to about 25 pounds, more than half of what driving the Prius (with two people) all the way to the Adirondacks entails. In fact, two people driving a 50-miles-per-gallon Prius to downtown and back (using a little over 1 gallon of gas) would

The author's home in winter.

Photo courtesy of Karl Coplan.

emit about the same amount of carbon dioxide (25 pounds per person). And my electric car would have lower emissions even if I drove it alone there and back and charged with the regular mix of electricity available on the New York City grid.

But I wasn't sure whether the Smart electric car had enough range to make it to midtown and back in the cold weather without a charge. Running the heat in cold weather really saps the battery and reduces the range; I can still make my daily commute to White Plains with no worries, but I did not want to have a dead car in the middle of the George Washington Bridge on the way home. However, once I realized that taking the bus and the train had a significant carbon impact, I decided to give the convenience of my personal automobile a try, at least for the Saturday sessions of the conference, when I would not have to deal with rush hour traffic. It's only a half-hour to midtown by car (without traffic), compared to nearly 2 hours by public transportation on a workday. To be safe, I looked up Manhattan garages with charging stations on my PlugShare phone app and headed for the Imperial Garage on West 54th Street. But when I got there, I discovered I needed a special account with the WattStation charger in the garage, and I was unable to set up an account online on the spot, so I took my chances. The Saturday environmental law session was a report on the recently concluded Paris negotiations implementing the Framework Convention on Climate Change. While some of the presenters celebrated the consensus on limiting global warming to 1.5°C and eliminating fossil fuel emissions by midcentury, others gloomily pointed out that the Paris agreement made no actual commitments to achieve the necessary reductions. At least I made it home after the session with 20 percent battery capacity remaining.

Since spring semester classes don't start until after Martin Luther King Day, I indulged myself with a working vacation at our Adirondack cabin during the second week of January. There was just enough snow to ski first thing in the morning before settling in for writing, case work, and class planning projects the rest of each day. I still had plenty of the firewood I had split in the fall, but to stretch it I cut down one of the standing dead red pines on the edge of our clearing; these standing stumps are usually well dried out, easy to split, and ready to burn right away. I used the electric chainsaw to cut the log into woodstove-size logs and charged the chainsaw batteries from the solar panels on our cabin. It's hard to regulate the heat sometimes in the cabin, and that weekend we had the cabin temperature well up into the 70s.

After classes started, I experimented with the heat at our home downstate. It was my goal to heat the house this year primarily with the fireplace-insert woodstove, using the seasoned deadwood from our backyard. I wanted to keep the gas furnace for backup, mostly, to keep the pipes from freezing if the woodstove couldn't keep up. So I set the thermostat for the furnace at 50 degrees and left it, and I kept running the stove to try to keep the temperature from dropping below 50 in the living room even when we were asleep under the down comforter or away at work. This was working pretty well, and I don't think the furnace was coming on at all as long as we were home to feed the woodstove last thing before going to sleep and in the morning before leaving for work. Most of the time, the temperature in the living room stayed in the low 60s while we were home in the evenings.

HOW TO CUT CARBON, SAVE MONEY, AND LOSE WEIGHT IN YOUR SLEEP WITHOUT DIETING OR EXERCISING

Setting the thermostat to 50 while we sleep is the only way we can keep the gas furnace from blowing our carbon budget. Keeping the thermostat down also saves us hundreds of dollars in our heating bill each winter.

Robin and I both grew up in houses that were cold at night. In Robin's case, it was good old Yankee frugality. In my case it was because my Dad never got around to fixing the radiator in my bedroom after it blew up when I was six. So turning the thermostat way down at night and snuggling under a thick down comforter is a comfort, not a hardship for us.

The fact is that, with a thick enough blanket, the human body can generate plenty of heat to stay warm and comfortable even in arctic temperatures. The more surprising fact is that your body will burn a significant number of calories to generate that heat. Researchers have been studying the role of something called "brown fat"—a kind of fat our bodies generate when we are exposed to temperatures below about 60 degrees. Instead of storing calories, brown fat burns calories to generate heat. Better to burn those calories to stay warm than to add them to your waistline!

Moreover, hugging your spouse close every night to stay warm is good for your marriage.

Sound too good to be true? Read the article about brown fat in *Scientific American*: "Supercharging Brown Fat to Battle Obesity: Why Turning Down the Thermostat Could Help Win the Battle of the Bulge," *Scientific American*, July 2014.

My wife Robin is very patient with me and my low-carbon ways. She also grew up in New Hampshire, works as a polar regions scientist, and is used to cold weather. Actually, some of our first dates were winter camping trips in the Adirondack Mountains. But even Robin drew the line at having company arrive while our woodstove was struggling to get the living room temperature back up into the 60s. This gave me a chance to see just how much gas, and carbon emissions, we needed to heat the house a little less frugally. On January 25, I left the gas thermostat at 50 (the heat never came on) and checked the gas meter: we used .5 ccf (ccf stands for "hundred cubic feet") in 24 hours (for the water heater and cooking). I then set the thermostat to warm the house up to 62 first thing in the morning, and again in the evening when Robin's polar research colleagues were coming to dinner. I checked the meter again: 24-hour gas use was about 2.8 ccf even with the heat on just in the morning and the evening. This works out to about 30 pounds of carbon dioxide emissions—equal to about a gallon and a half of gasoline, or the equivalent of driving a car alone to work that day. The temperature range that day was about normal for late January: a low of 24 and a high of 36 (the previous night, using the woodstove alone, was much colder).

We also had an old-fashioned winter storm this month, with about 18 inches of snow in our driveway on Saturday, January 23. New York City nearly broke its all-time snowstorm record that day, with 26 inches falling in Central Park. We let the snow pile up all day and enjoyed a snow day in front of our fireplace, then went cross-country skiing in a marsh nearby in the late afternoon as the storm wound down. As we skied back down our driveway, a creature with a fuzzy tan tail silently bounded away through the bushes. I later identified the visitor as a bobcat. Sunday morning, we shoveled out the driveway by hand together (with our usual debate about the most efficacious way of shoveling out a driveway). Shoveling snow, like chopping wood, is a nice way to get some fresh air and exercise outdoors, but I suppose if I ever get too frail to shovel by hand, I can always buy an electric snowblower without increasing my carbon footprint. Monday morning, I worried whether my little electric car would make it up the snowy driveway, but I had no trouble in the driveway and didn't have any problems handling the cleared roads on the way to work.

My total carbon footprint for the month was about 500 pounds of CO_2, which keeps me well on track to stay within my 4-ton annual carbon budget. The big-ticket items for the month were the gas for our trips upstate and natural gas, which was not a surprise since even with the woodstove the water heater runs longer in the winter and the furnace still comes on sometimes.

6

GOING ON A CARBON DIET TO SAVE THE PLANET

If you are with me so far, I hope I've convinced you that, first, everyone has an ethical obligation not to exceed a sustainable individual carbon footprint; second, that to be "sustainable," an individual carbon footprint cannot exceed one's individual share of the roughly 15 gigatons per year of greenhouse gases the globe can afford to emit between now and a mid-century total phaseout of fossil fuels; and third, that for a middle-income individual in the United States, a carbon footprint of about 4 tons of direct carbon dioxide impacts is sustainable based on existing patterns of wealth and consumption. Since the average American currently is responsible for about two to three times that level of emissions, this means that living sustainably will involve tightening your carbon belt substantially.

Belt-tightening might not be a bad metaphor, since the lifestyle changes necessary to live sustainably are analogous to the lifestyle changes involved in going on a diet to lose weight. Just as a basic calorie-limited diet involves trade-offs between small quantities of high-caloric treats and larger quantities of healthier low-calorie foods, a sustainable carbon footprint involves trade-offs between high-carbon "sweets" (like flying to the Caribbean for a winter vacation) and lower-carbon basics (like getting to work every day in a gas-powered car). Just as you can look up the calorie count in the foods you eat and make intelligent choices to stay within a low-calorie goal, you can look up the carbon count of all the life choices you make to stay within a low-carbon goal. Living "sustainably" means setting an overall carbon limit that, collectively, would maintain an acceptable global climate, just as maintaining a healthy weight might require setting a daily caloric limit.

A MEDITERRANEAN DIET APPROACH TO REDUCING YOUR CLIMATE FOOTPRINT

Going on a carbon diet doesn't mean giving up all luxuries, any more than going on a healthy diet doesn't mean giving up all rich foods. It just requires some moderation and smart choices. I like to think of it as the Mediterranean diet approach to your personal carbon footprint. About fifteen years ago I had a high cholesterol and triglycerides count at my annual checkup. I had never paid much attention to my diet before that, since I was one of those blessed people who could eat pretty much anything without gaining much weight. Three cheeseburgers for lunch? Why not?

Back then, my doctor automatically prescribed statin drugs for every patient with high cholesterol. I resisted. I was too young to have to take meds every day for the rest of my life. But to me, food is one of the essential pleasures in life, and I didn't want to swear off meat, or any other food group, completely. Robin handed me Nancy Harmon Jenkins's *Mediterranean Diet Cookbook*, which was already sitting on our shelf, and it pointed to a way out of my predicament: a book full of delicious-sounding food, all somehow associated with very low incidence of heart disease.

Jenkins quotes this description of the Mediterranean diet from a 1993 Harvard Medical School conference:

> Plentiful fruits, vegetables, legumes and grains; olive oil as the principal fat; lean red meat consumed only a few times a month or somewhat more often in very small portions; low to moderate consumption of other foods from animal sources such as dairy products (especially cheese and yogurt), fish, and poultry; and moderate consumption of wine (primarily at meals).

Sounds good to me! As Jenkins puts it, "A diet of Mediterranean dishes . . . will be healthful and satisfying but never austere." Best of all, this diet didn't require me to totally give up meat or carbs or fats or anything else—just moderate my consumption of some foods. An article on the National Institutes of Health website summarizes the diet this way: "In a somewhat reductionist approach, the traditional Mediterranean diet can be considered as a mainly, but not dogmatically, exclusive plant-based dietary pattern."

Not dogmatically vegetarian! The Mediterranean diet, in which red meat is either a garnish in small quantities or a rare treat at a celebration, is pretty good for the climate too, since beef and lamb have about the highest climate footprint among food choices (unfortunately, dairy products have big footprints, too). But part of the Mediterranean diet and culture also includes "feast days"—usually religious celebrations when people indulge in the rich food that is not part of their daily fare—so you can still enjoy these rich foods as a treat.

This means that a beautiful and savory minestrone soup is good for my health and for the climate. It's not vegetarian; you start by frying a little pancetta in the soup pot, and adding beef stock enriches the flavor. I am not dogmatic about vegetarianism. Or my "climatarianism," for that matter. When climate-conscious friends chide me for eating some free-range organic beef about once a month (13 pounds of CO_2e for an 8-ounce portion), I smile and resist comparing my meat eating to their jet travel (2,000 pounds of CO_2e for a typical round-trip flight). I might even get on a plane myself for a really good reason, as long as it fits in my carbon budget. We all deserve our occasional feast day.

After about three months of a Mediterranean diet, together with adding daily exercise to my routine, my cholesterol was back in the normal range. Collectively, tackling our carbon diet can have similar benefits for the health of the planet, though not as quickly.

SETTING A PERSONAL CARBON BUDGET

Living with a sustainable carbon footprint is also like setting a personal financial budget. You only have so much cash to spend each month, so you have to set money aside for the absolute necessities (mortgage or rent, electricity and heat, food, child care, transportation to work) and hopefully have money left over at the end of the month for luxuries that enrich your life: travel, recreation, movies, shows, sports events, gifts to charities and friends. If you plan well and live within your means, you will have money left over for these activities. If you don't plan well and try to live above your means (with a bigger house than you can really afford, for example), there will not be enough money left over at the end of the month for fun.

Like following a calorie-counting diet, setting a personal carbon budget means picking an overall limit on your carbon footprint and figuring out how to live within that limit. But a carbon budget is more like a financial budget in many ways: it takes more planning and mindfulness to accomplish. Calorie counting can be done on a daily basis, and if you blow your calorie goal today, you can always try to resume your diet tomorrow. However, trying to set a daily carbon budget won't work because your carbon footprint for the basic necessities of life is likely to vary over the year. For example, the carbon impact of heating your house will be much higher in February than in April. So you have to plan ahead to figure out how much extra carbon footprint you need for high-carbon months of the year. This is not unlike planning ahead in a financial budget to make sure you have enough money for big expenses that come up regularly but not every month, like your property tax bill or your car insurance renewal. You probably have some fun expenses, too, like the money you spend on vacation travel or holiday gifts. So, unlike the diet that you keep track of on a daily basis, your carbon budget is probably better allocated over the entire twelve-month cycle of seasons; you might keep track on a monthly basis to see what's left in your carbon account, but you don't know whether you are keeping to your carbon budget unless you look back and see how your footprint for the last twelve months adds up. As the last chapter has suggested, this twelve-month budget should probably be in the range of 4 to 5 tons to be considered sustainable for a middle-class American.

You can think of it this way: if you blow your financial budget and spend more in a month than your take-home pay, you will soon use up your savings. Your checks will bounce and you will start running up a credit card balance. In a way, that is what we are doing collectively on both a national and a global basis, since we are emitting greenhouse gases at a rate far in excess of what the global ecosystem can process. Our greenhouse checks are bouncing, and our climate creditors will ultimately be very unforgiving! But the climate credit system works slowly. Since we are collectively overspending our greenhouse gas account, we are running up a massive carbon credit card debt—one that our children will have to pay for. If we are responsible parents and stewards of the planet, we will avoid running that credit card debt up, even though it is someone else who is likely to pay for it. Especially since it is someone else who will have to pay for it.

Sticking to a carbon budget requires some dedication and willpower; in this sense, it is more like a weight control diet than a cash budget. If you blow your cash budget for the month and overdraw your account, your bank will eventually stop honoring your checks and start declining credit card charges, which will stop you from spending any more of the money that you don't have. When you overdraw your carbon budget, nothing happens except that you know you are contributing disproportionately to a global calamity. This works more like a weight loss diet than a budget: when you blow your diet for the day, restaurants don't stop serving you food but you know that weight gain will follow eventually. Sticking to a carbon budget, like sticking to a diet, will not be easy, especially when all your friends are asking you why you won't share that second slice of birthday cake or that airline trip to a "destination wedding" with them.

WHAT ABOUT OFFSETS INSTEAD OF PERSONAL REDUCTIONS?

If your goal is to model a sustainable carbon footprint, then you should be highly skeptical of offsets as a tool to reduce your climate footprint. The idea of an offset—paying someone else to reduce his or her carbon emissions or to engage in activities that reduce or sequester carbon—is not consistent with adjusting your own consumption choices. Many offsets are collecting money for things that will have to be done anyway to address climate change, or they depend on questionable science or predictions about the future. If your offset consists of helping to finance a solar power project in a developing country, for example, chances are that that solar power plant would have been built anyway during the carbon "payback" period you are taking credit for, as the world must convert to zero carbon energy sources in just a few decades. If your offset consists of paying someone to promise to plant and preserve a forest somewhere in the world, then the offset depends on uncertain predictions about the sequestering potential of forests in different climates, as well as the enforceability of that promise decades in the future when the political and legal regimes may have changed and demand for biomass for energy may be high. In the case of either the power plant or the forest preservation, your carbon "offset" will occur decades into the future, while your carbon emissions occur over the course of days (or hours, in the case of airline flights). If you are paying less than the social cost of 1 ton of carbon (currently about $50 per ton, but increasing exponentially), then the offset does not even serve the economic function of discouraging activities that cause more climate harm than they are worth.

I don't use offsets because my goal is to reduce my footprint to sustainable levels and model the low-carbon good life. But some personal footprint reduction measures are very similar to offsets, such as purchasing renewable energy credits to offset your home electricity consumption in a market where green power is not available. For me, to be creditable, an offset should "pay back" your carbon emissions within one year and be priced at no less than the $50 per ton social cost of carbon.

ENVIRONMENTALISM AS AN ACCOUNTING SYSTEM

When I teach my course in environmental law, I start out by suggesting that all environmental regulation is just a kind of accounting system. Environmental regulation is a matter of getting people to incorporate environmental costs into their bottom line and to consider those costs in making decisions. Since the market economy fails to include environmental costs, regulation is necessary to bring those costs into the market accounting system. For example, if a developer sees she can make $2 million by destroying a wetland and building a shopping mall, that developer only sees profit, even if the loss of the wetland causes $3 million in flooding damages and losses to a productive fishery. Environmental regulation makes society consider the environmental losses built into private profit-making activities and adjust the incentives.

Since, just like that developer, we don't have to pay for the climate losses caused by our carbon-emitting activities, addressing climate change is going to require some sort of accounting on an individual as well as a national basis. Sticking to a carbon budget is a sort of accounting problem, then. Counting calories in a diet is fairly easy. You can keep track of what you eat in a given day and read the boxes for your prepared foods to see how many calories they contain. When you eat at a restaurant, counting calories may be a greater challenge, but many cities now require that restaurants also list the calorie content of their meals. Home-prepared meals may be even more challenging to track, since you would have to look up the calorie content of each raw ingredient, multiply by the amount used, and divide by the number of portions. But at least you know how much of each food you actually ate and what should count against your diet.

Counting carbon impacts is more complicated. Sure, some carbon impacts are easy to keep track of: you know, when you buy 10 gallons of gas and burn it all up by driving your car all alone, that you are responsible for 10 gallons' worth of carbon impacts. That's about 200 pounds, since a gallon of gasoline represents about 20 pounds of carbon dioxide. But what about when you ride the bus to work? Some people would argue that "well, that bus is driving on that route anyway, so the carbon impact of my taking the bus is zero." Following that same reasoning, you might argue that "well, that plane is flying from New York to Los Angeles anyway, so the carbon impact of my flight is zero." Under this argument, we can all be "free riders" on someone else's carbon footprint. I have heard people make these arguments, but I don't believe them. Few vegetarians would accept the argument that "well, that calf was going to be slaughtered for veal anyway, so when I eat veal I am not actually causing cruel treatment of any particular animal." The problem with these arguments is that they are a way of avoiding personal responsibility for the collective impact of our choices. Of course that plane will fly to Los Angeles whether I go or not, but if everyone on that plane decided the trip was not worth the climate change impacts, then the plane would be empty and the emissions would not occur.

This sort of collective action problem underlies the fundamental challenge of incorporating environmental values into market economy decision making. Biologist Garrett Hardin famously described this challenge in his influential 1968 essay, "The Tragedy of the Commons." Hardin described a parable in which several sheepherders share a common pasture, owned by the public. Each sheepherder has an incentive to keep adding sheep to his flock because he reaps the entire benefit of selling one more sheep, while the cost of each additional sheep—overgrazing of the pasture—is shared by all of the sheepherders. Each sheepherder acts rationally for his own profit, adding sheep without limit, until the entire common pasture is destroyed by overgrazing and all of the sheepherders lose their livelihood. Individual profit maximization thus leads to environmental and economic catastrophe for all.

The tragedy of the commons is an example of what economists call "market failure"—the failure of an unregulated market to maximize social and economic goods in the way that efficient market theory expects. Economists accept that, in the case of market failure, government intervention is appropriate to correct the failure and restore the public welfare. It is easy to see how the market "fails" in Hardin's example: each extra sheep has a true cost, but the owner of the sheep doesn't have to pay the full cost. These uncaptured costs are considered "externalities": they are left "outside" of the profit-and-loss statement for the business (the sheepherder) even though they are a true cost to society.

The challenge of environmental policy generally is to "internalize" these external costs so that they are part of the decision-making process for each actor in the environmental economy. It's easy to see how the tragedy of the commons applies to natural resource issues like grazing lands and fishing rights, where individual profit maximization leads to destruction of the entire common resource that generates the profits in the first place. But the tragedy of the commons paradigm applies to environmental pollution as well, where one polluter might be able to dump its wastes in a river without observable effect, but one hundred polluters will destroy the river ecosystem, fishery, and drinking water resource for everyone. Absent regulation, polluters don't have to pay for the cost of environmental degradation, so those costs don't show up on the polluter's accounting of profit and loss. Environmental regulation seeks to make the polluters responsible for the degradation they cause, either by prohibiting the harm or by making the polluter pay for the harm. If polluters have to pay for their harm, they no longer reap false profits at the expense of public environmental resources. In this way, environmental regulation is simply an accounting system—an attempt to quantify (or eliminate) the environmental harms and to capture them in the profit-and-loss statement of whoever causes those harms.

Of course, climate change is a tragedy of the commons involving a "commons" that consists of the entire atmosphere of the planet together with all of the natural systems that process and recycle carbon dioxide back into carbon through photosynthesis. Individual people and businesses that add to this problem don't have to pay for the collective harm, so their activities seem cost-free in the market economy. Bringing these costs home to everyone who contributes to the problem involves enormous accounting challenges because it is difficult to attribute common costs to their true owner. Living sustainably on an individual carbon budget requires some accounting choices as well.

For example, let's consider how we might come up with the carbon impacts of taking a bus. If you take the approach that "that bus was going that way anyway, so my impact is zero," you are essentially treating the bus ride as a commons: no one "owns" those carbon impacts. Perhaps more accurately, the carbon impact of operating the bus is a "common" cost for society and is part of our shared indirect carbon impact, like the carbon impact of keeping the lights on at city hall. But treating the bus (or the jet plane) as a "common" resource doesn't address the tragedy of the commons problem; it simply perpetuates it in a new context. If you were to assume that someone has to take responsibility for (or "pay") the carbon costs of the bus, you might decide that the fairest way to allocate that cost is to charge the riders of the bus, since they get the most personal benefit from the bus trip. You might consider that, if we imposed a dollar price on carbon emissions (through taxes on fossil fuels or tradable rationing certificates for fossil fuels), the bus riders would most likely have to pay for those emissions taxes through higher fares.

But how should I account for my share of carbon from that bus? You could take the fuel economy of the bus, count the number of riders each time you get on, and calculate your share. But what if you get on the bus and it is practically empty? Do you then have to take responsibility for all of the emissions of a 5-miles-per-gallon bus? Driving alone in the biggest gas-guzzling SUV would be a better choice. Or you could take the average bus fuel economy and the average bus ridership and come up with a workable rate in pounds of CO_2 per passenger mile.

So keeping track of a carbon budget is going to involve some accounting issues. Reasonable people may differ. My rule of thumb is that if you are paying for it, the emissions belong on your carbon tab: that plane that is flying to Los Angeles anyway is on your carbon budget unless the airline says, sure, board for free, because that plane was going to Los Angeles anyway. I use published carbon emissions rates where available for calculations such as carbon emissions per passenger mile for intercity and city bus travel, rail travel, natural gas per ccf (hundred cubic feet), and gas and diesel per gallon (see table 6.1).

Online carbon calculators are very useful, too—with a caveat. The best calculators allow you to enter your actual electricity, fuel oil, and natural gas consumption for household utilities, rather than simply giving you an average based on the size of your house. And I have still not found any carbon calculator that is consistent in all aspects with published scientific studies on the impacts of the underlying activities. For example, carbonfootprint.com does a good job with air travel, natural gas usage, automobile mileage, and bus travel, but it vastly understates

TABLE 6.1 Some Rough Greenhouse Gas Conversion Factors

EMISSIONS SOURCE	POUNDS OF CO_2E (APPROX.)
One gallon of gas or diesel	20
One ccf of natural gas	12
One pound of beef or lamb	27
One mile in a train	0.35
One mile on a bus	0.13
One kWh of electricity	1

Note: ccf = 100 cubic feet natural gas; CO_2e of 1 kWh varies from 0.3 to 1.7, depending on region.

the impacts of rail travel—by a factor of 10. These online calculators are constantly being updated and are becoming more thorough, and many now include "indirect" impacts, like life cycle impacts of food and consumer goods, that previously were untracked because they were part of an individual's indirect footprint (see chapter 5 for a definition of "indirect" footprint).

Since I was not satisfied with any of the online carbon calculators, I put together a spreadsheet to keep track of my own daily carbon-emitting activities, based on the best emissions factors I was able to find. I kept it simple, tracking only the most significant and most easily calculated items, like fossil fuel consumption and high-impact foods like beef and lamb.

We'll take a look at the specific climate impacts of various individual activities next.

REFERENCES

Coase, R. H. "The Problem of Social Cost." *Journal of Law and Economics* 3 (1960): 1–44.
Hardin, Garrett. "The Tragedy of the Commons." *Science* 162, no. 3859 (1968): 1243–1248.
Harmon Jenkins, Nancy. *The Mediterranean Diet Cookbook*. New York: Bantam, 1994.
Stiglitz, Joseph. "Regulation and Failure." In *New Perspectives on Regulation*, ed. David Moss and John Cisternino, 11–24. Cambridge, MA: The Tobin Project, 2009.

CARBON DIARY

February 2016

On the first of the month, Robin gave a lecture on her polar research projects at the Explorer's Club in New York City, with a postlecture dinner invitation nearby. By now I had figured out that the lowest-carbon way to get into the city for the evening was to drive my little electric Smart car, but would it have the range to go from my office in White Plains to New York City and then back to Rockland County? The round trip would be a total of about 60 miles.

To be safe, I researched garages with charging stations ahead of time, activated an account with ChargePoint, called the garage on East 80th Street ahead of time to confirm availability, and headed off. When I got to the garage, however, and asked the attendant about charging the car, he said there was no charging station at that location, even though my ChargePoint app had steered me to the door. The attendant told me that the charging station was at the garage across the street, run by the same company. So I drove to the garage across the street, where that attendant informed me that there was no charging station there, but that the charging station was at the first garage.

Since I was already running late for the reception before the lecture and it was going to be a 10-block walk to the lecture from 80th Street, and returning to the first garage meant either driving the wrong way on 80th Street or going all the way around the block again, I gave up, drove to 70th Street, parked the car in an insanely expensive garage full of Land Rovers and Lamborghinis, and took my chances on getting home without charging. Strike two for New York City electric car charging stations. We ended up making it home without any trouble. The round trip still cost

The author's winter cabin.

Photo courtesy of Karl Coplan.

me 2 pounds of carbon for topping off with the not-necessarily-renewable electricity at my office.

My little Smart car is turning out to be a perfect way to get around with zero direct carbon impact. Its range varies with the temperature, but in general I can get between 50 and 100 miles between charges. That was enough, this month, to take Robin out to the Banff Mountain Film Festival at the classic art deco Lafayette Theatre in Suffern, about 15 miles from our house, even after driving to work and back the same day. And it was enough to take Robin and our dog Lupie along for a hike up the Pine Meadow Gorge in Harriman State Park, about a 65-mile round trip (including detours to Haverstraw to check on our wintering sailboat and to Nyack to shop for a dinner party). I even had enough charge on the car to drive one of our dinner guests across the river to the Tarrytown train station afterwards.

Despite being another mild month overall, President's Weekend featured a record cold snap in the Northeast; temperatures at our house in

West Nyack dropped to 8 below zero that Saturday night. We weren't there to see how the woodstove would cope with the low temperatures, however, since we went upstate to our cabin in the Adirondacks. I left the thermostat at home set to 48 and shut off the main water valve at the wellhead just in case the pipes in the cellar froze. When we arrived at our cabin, the interior temperature was 18 degrees, so we started our woodstove and headed off to Garnet Hill Lodge for a day-early Valentine's dinner out. With the woodstove cranked up, we had the cabin up to a comfortable 70 degrees despite the subzero daytime temperatures outside.

Three days later, it was pouring rain, even in the north woods. When we got back to our home downstate, everything was in order. No pipes had frozen. I was worried that when I totaled the natural gas carbon impacts up for the month, I would pay dearly for leaving the house heated with gas during the cold snap, but the total natural gas carbon footprint for the month was about 200 pounds—less than January's total. My total carbon bill for the month was 353 pounds. This included the round trip to the Adirondacks in the Prius as well as two beef meals that I could not resist.

With the end of February, I have been tallying up my monthly carbon footprint for six months now. My footprint for half a year—including the three coldest months of the year—is about 2,380 pounds. So I am well within my 4-ton carbon budget for twelve months, meaning I can plan for some fun this summer without blowing my budget!

7

SURPRISING CARBON IMPACT COMPARISONS

If You Are Only Going to Sweat One Kind of Stuff, Sweat Big Stuff, Not Small Stuff

If you are reading this book, you have probably already taken some steps to reduce your carbon footprint. You drive a hybrid car, or maybe even an electric. You long ago got rid of your incandescent light bulbs and replaced them with compact fluorescents and LEDs. You take public transportation to work if possible. You shop at the local farmers' market in season. But you probably get on a jet plane several times a year. And you figure you have done your part, and that addressing global warming will simply require social or regulatory changes that will get the rest of the population with the program you have already adopted.

But you have probably not performed the sort of accounting I talked about in the last chapter, so you may not know how those discrete carbon-reducing measures stack up compared to the average American's footprint. People tend to look for just one or two changes to address a problem. Even worse, due to virtue signaling, people who have taken virtuous steps in one visible area of their carbon footprint, like food consumption, often feel justified in consuming more resources in other areas. Without knowing how much of a difference these steps make, we risk taking purely symbolic measures that make no real difference to our contribution to the problem, or we rely on symbolic reductions to justify outsized consumption impacts. It is like going on a diet without knowing how many calories are contained in different foods: we might then choose to give up broccoli (never really liked it anyway) while continuing to eat cheeseburgers. Government and environmental programs compound this problem by emphasizing the "small ticket" reductions, like trading in your incandescent light bulbs and your conventional automobile, and ignoring the big-ticket items like commuting distance and air travel.

So let's take a look at some common individual activities, their climate impacts, and the effects of common footprint reductions. Most environmental organizations advise commonsense measures like turning your thermostat down a few degrees in the winter, driving a more fuel-efficient hybrid car, recycling your garbage, and buying locally produced food. How much of a difference do these measures really make in the context of the average American carbon footprint of about 20 tons of carbon dioxide? How do the carbon emissions saved by taking public transportation to work compare to the savings from foregoing air travel and driving to a family vacation closer to home, or giving up red meat versus buying all your vegetables locally? How significant are measures like changing out all of our incandescent bulbs compared to the average American carbon footprint? Some of the results are surprising . . .

DRIVING AN SUV CAN BE BETTER FOR THE PLANET THAN FLYING

Dedicated environmentalists wouldn't dream of driving a gas-guzzling sport utility vehicle, but they usually don't think twice about packing the family onto a jet plane for a vacation close to nature. So what's worse for the climate, packing a family of four into an SUV and driving a 2,000-mile round trip to Disneyland, or packing that same family of four onto a jet plane for an ecotour in Costa Rica?

It turns out the family in the SUV gets credit as good planetary stewards, not the international travelers. Assuming that the SUV (such as a Chevy Suburban) gets 20 miles per gallon on the highway, that 2,000-mile round trip from New York to Orlando, Florida, will set the planet back by about 1 ton of CO_2—or about 500 pounds of CO_2 per passenger. Each passenger on a round-trip jet flight from New York to Costa Rica will rack up four times as much CO_2 emissions. Not only that, but that one trip will suck up about a quarter of my 4-ton annual carbon budget, not leaving enough for stuff like getting to work and heating the house. You wouldn't spend one-quarter of your household income on the family vacation, would you?

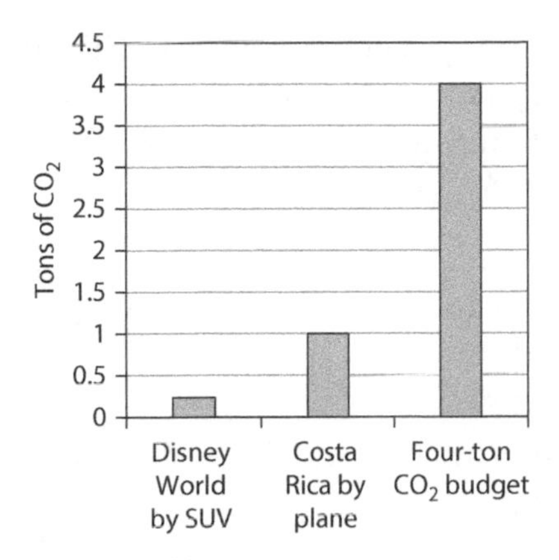

Carbon footprints for trips to Disney World and Costa Rica.

So if you have been an activist running around pasting "Climate Violator" citations onto SUVs while jetting off to exotic ecosystems on holiday, you might want to go back and leave some apology notes.

BEING A LOCAVORE IS GOOD, BUT BEING A VEGETARIAN IS BETTER

Many dedicated environmentalists are dedicated locavores as well: they eat locally grown food whenever possible. There are many great benefits of being a locavore—support for family farms, the resilience and community benefits of community agriculture, and the environmental benefits of low-impact and organic farming methods. But the climate impacts of transporting food even long distances to market are really minimal. Freight transport is much more efficient than people transport; you may remember hearing advertisments by the CSX railroad claiming that they can transport a ton of cargo 400 miles with just one gallon of fuel. A study by Pennsylvania State University concluded that the annual carbon footprint of transporting the vegetables consumed by the average household was just 0.15 ton, or about one-twentieth of a ton (0.05 ton) per person in a three-member household. You would emit more carbon dioxide by driving 4 miles round trip to the farmers' market once a week in the average car.

On the other hand, beef and lamb have a huge footprint, primarily because of the methane and nitrous oxide emissions they produce.

One pound of beef sends about 28 pounds of CO_2e into the atmosphere. And the average American consumes about 70 pounds of red meat a year—for a total of 1 ton of carbon dioxide emissions.

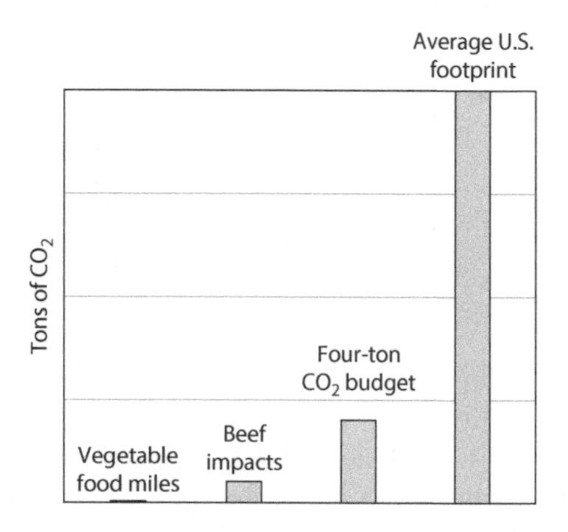

Carbon footprints for locavorism.

So reducing your beef consumption is a far, far more effective thing to do than going to the local farmers' market. And that 0.05 ton of carbon impacts for transporting foods long distances to the supermarket is not going to make a dent in the average American footprint.

LOCATOURISM BEATS LOCAVOURISM EVEN ON THE TRAIN

That vacation trip to Costa Rica cost one-quarter of an individual annual carbon budget. How about taking the train across the country instead? Greenies love trains; they seem to be the epitome of slow travel, shared community, and low impact—with a little bit of nostalgia thrown in. Surely, train travel must be climate friendly, too.

As it turns out, while rail travel has lower impacts than air travel, especially for shorter distances, it is not carbon neutral, not by any means. Hard figures are hard to come by. Amtrak's website coyly says that the carbon impacts of rail travel are "less" than the impacts of flying. But whenever I try to get the real truth about the carbon impacts of rail travel and tote up the numbers for my personal carbon budget, all of my illusions are shattered and I am reminded that the only kind of long-distance travel that is really sustainable is ocean sailing, or bicycling, or maybe being packed into an intercity bus, or sharing a road trip with two other people in a Prius.

A 2008 Union of Concerned Scientists (UCS) report on CO_2 impacts of travel calculates Amtrak's impacts at about 0.37 pound per passenger mile for the electrified Northeast Corridor, or about 0.45 pound per passenger mile for the rest of the rail network. When I worked out the impacts of my train trip to Wilmington, North Carolina, for the Waterkeeper conference in 2016, using the UCS numbers, I got a total of 209 pounds of CO_2, or about 0.1 ton of CO_2. I ended up flying back to New York from Charlotte, North Carolina (I had to teach my class Thursday afternoon). When I plug that flight into carbonfootprint.com, it shows emissions of 0.12 metric ton of CO_2, or about 0.13 English ton of CO_2. So taking the train might only be about 25 percent better than flying. And lots of long-distance travel by train or plane is not ever going to be consistent with a sustainable carbon footprint.

Things get even more complicated, though, since air travel involves impacts beyond just those caused by burning fuel: there's also water vapor and nitrogen oxides at high altitudes, which have potent greenhouse gas effects. Some studies suggest that the true air travel impacts are double those implied based on fuel use—as much as 500 grams per passenger mile, or about 1 pound per passenger mile. Amtrak would then cause only half the emissions of flying. But one-half of a very large number is still a large number. In other words, cutting the total number of miles traveled is more important for making a significant reduction in your footprint.

MY ELECTRIC CAR BEATS YOUR ELECTRIC TRAIN

I have a friend, a dedicated environmentalist, who commutes 80 miles a day via the Metro North commuter trains to get to work and back. He doesn't calculate his carbon footprint, but he takes credit for his green commute; after all, every environmental group tells you that public transit is a big part of the solution to climate change. Several colleagues of mine at Pace Law School in White Plains commute back from New York City to our White Plains campus; more than one of them has suggested that his urban existence and public transit commute must have a lower carbon footprint than my suburban lifestyle.

But when I look at the actual numbers, commuter rail does not seem to come out ahead of even a solo passenger electric car. According to the Metropolitan Transportation Authority's 2012 sustainability report, the Metro North railroad has a direct carbon footprint of 0.27 pound per passenger mile. When I look at my electric Smart car, which gets about 3 miles per kilowatt-hour of electricity (60-mile range on an 18-kilowatt-hour battery), the carbon footprint works out to about 0.25 pound per passenger mile, based on an EPA estimate of 0.75 pound of CO_2 per kilowatt-hour of electricity in Westchester County. So driving alone in an electric car is actually a slightly lower carbon footprint than taking the commuter train! Even two people in a 50-miles-per-gallon hybrid can beat the train: a two-passenger hybrid vehicle works out to about 0.2 pound of CO_2 per passenger mile (20 pounds of CO_2 per gallon divided by 100 person-miles-per-gallon for two people in a hybrid).

Public transit plays a big role in reducing congestion and moving large numbers of people to work every day; it prevents the gridlock that would result if everyone tried to drive to work. Moreover, the New York City subway system does much better at moving large numbers of people

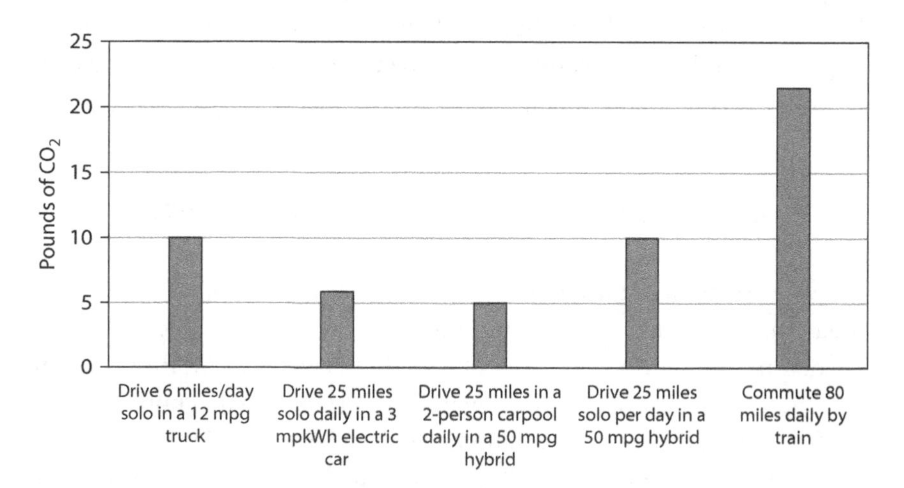

Commuter carbon footprints.

with a small footprint than commuter rail—just 0.12 pound of CO_2 per passenger mile. So a subway commute beats out any sort of car. New York City buses, on the other hand, are not very efficient. In the long run, electrified commuter rail has the advantage of potentially being powered by 100 percent renewable energy. But how far you commute matters more than how you get to your destination. Greenies in the metropolis, with their long commutes, may have less carbon credibility than the middle American who drives her truck 6 miles to work and back each day.

THE BIG PICTURE ON THE BIG-TICKET ITEMS

If we are going to achieve the carbon reductions necessary to save our climate system, we will have to attack the big-foot parts of our individual footprints, not just the toe prints. Here is a chart that helps put the carbon footprint of some of our lifestyle choices in context:

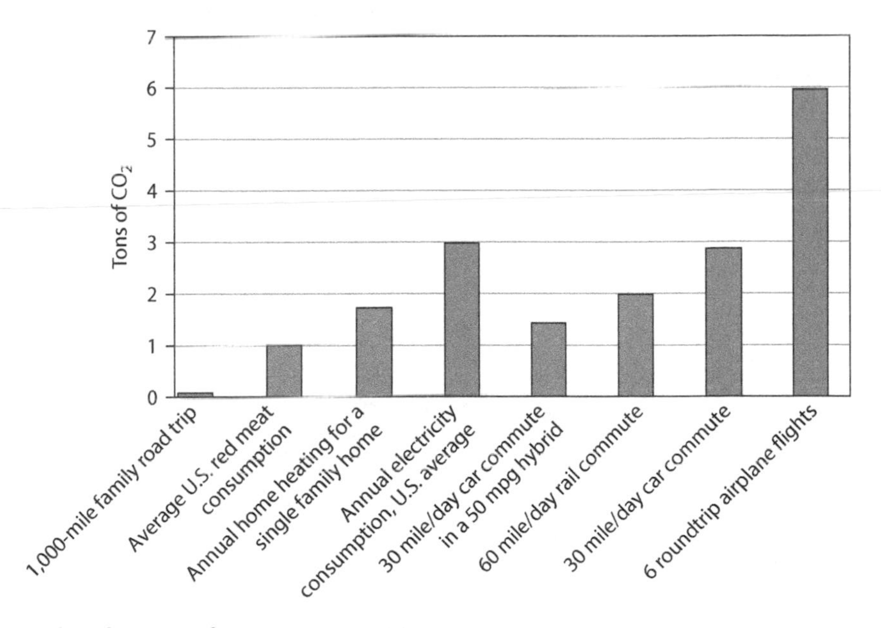

Carbon footprints for common activities.

Data sources include "Shrink Your Housing Footprint," ShrinkThatFootprint.com, http://shrinkthatfootprint.com/shrink-your-housing-footprint.

What this chart reveals is that home lighting and heating, transportation, and food make up a huge part of a typical American's carbon footprint. Just heating and lighting your house will quickly consume a 4-ton annual carbon budget, leaving nothing for commuting to work, food, or other travel. So a plan to keep to a carbon budget is going to have to address these Big Four big-ticket items first, in order to leave some budget left over for the things that enrich life, like good food and travel to interesting places. We turn to the big-ticket items next.

REFERENCES

Metropolitan Transportation Authority. *An Average MTA Trip Saves over 10 Pounds of Greenhouse Gas Emissions.* April 2012. http://web.mta.info/sustainability/pdf/2012Report.pdf.

Union of Concerned Scientists. *Getting There Greener: The Guide to Your Lower-Carbon Vacation.* 2008. https://www.ucsusa.org/sites/default/files/legacy/assets/documents/clean_vehicles/greentravel_report.pdf.

Weber, Christopher L., and H. Scott Matthews. "Food-Miles and the Relative Climate Impacts of Food Choices in the United States." *Environmental Science and Technology* 42, no. 10 (April 2008): 3508–3513.

CARBON DIARY

March 2016

Most winters for nearly two decades now, my son and I have arranged to get together for a weekend ski mountaineering and camping trip in the Adirondack Mountains. Now that Justin is an adult, it gets harder and harder to find a weekend when it works for both of us. This year, we settled in advance on the first weekend in March. Justin drove up from his home in Harrisburg, Pennsylvania, on Thursday, and on Friday we drove together up to our cabin in North River. This was Justin's first visit to the cabin. It had recently rained heavily on top of a thin coating of snow, and everything was a glaze of ice. When we arrived, it was colder inside the cabin than outside, so we fired up the woodstove, took a walk around the trails on the property to warm our bodies, and went to the lodge for dinner while the cabin warmed up a little.

There was too little snow in the Adirondacks for backcountry skiing that weekend, but the 3 to 4 inches of new snow on the balsams and red spruce made for a beautiful backwoods hike, and with the spikes on our boots, we made an easy climb of New York's second-highest peak, Algonquin. Saturday, March 5, was a perfect day for a winter ascent—light winds, clear skies, fresh snow—and the Algonquin summit was crowded with hikers. I was pretty sure I could pick out the ridge where our cabin is from the summit. The temperature went down to zero at our campsite overnight, but I had long ago learned that as long as you stay in your winter sleeping bag you can sleep quite comfortably. You just have to get moving fast in the morning, before your toes freeze up.

I continued to test the range on my electric Smart two-seater car. One Saturday, we were invited to a traditional rice-feeding celebration for

A view from Algonquin Peak.

Photo courtesy of Karl Coplan.

the infant child of one of Robin's colleagues, who is Nepalese. We didn't have to drive to Nepal, but Astoria, Queens, was at the edge of the round-trip range. Worse still, my iPhone navigator gave us bum directions, adding 3 miles to the trip. But our backup plan was that if we had to we could leave the car at Robin's lab on our way home, 10 miles short of our house, and figure out how to charge it. On the drive home in the rain, we even worried about whether running the windshield wiper would reduce the range. Luckily, the Smart car came through again.

We also traveled to Harrisburg to visit Justin and his girlfriend. Robin traveled to DC that week with our college-age daughter, Beryl, so I took the train to Harrisburg from New York and met everyone there. Once again, public transportation had a more significant carbon impact than driving the Prius: my carbon footprint for the New Jersey Transit train to Newark and then for Amtrak from Newark to Harrisburg was 70 pounds—nearly twice my share of the carbon footprint of driving home with Robin in the Prius, which is 40 pounds. This was partly due to the circuitous route necessitated by public transportation (Newark is not really on the way from our house to Harrisburg), but I am also learning to my surprise that public transportation is generally less carbon friendly than carpooling in a hybrid.

The rest of March was unseasonably mild, and this winter is likely to be remembered as the winter that wasn't. Meteorologists say that this was due to an unusually strong El Niño weather pattern in the Pacific, but the climate news this month was full of headlines that February broke global records for the warmest month on the planet since climate records started being kept, and that pattern continued into March. Robins and blackbirds were establishing their worming grounds and roosts right from the first week. Daffodils bloomed in our backyard by the middle of the month, and the red and gold of the leaf buds seemed to spring up about a month early. To our great pleasure, a pair of cooper's hawks moved into a tall tree in our woods during the third week of the month.

I don't think our gas-fired furnace came on once this month. The woodstove kept our living room warm enough to keep the furnace off, and our puffy down comforter kept us plenty warm in our bed at night. First thing each morning, before we went for our run, I started a brisk fire in the woodstove to warm our living room back up again. As a result, when I toted up my carbon bill for the month, I was pleased to see that our natural gas use was down to the lowest it had been since October and was no longer the big-ticket item for the month—gas for car travel was. Total carbon emissions for the month were 403 pounds.

8

GRAPPLING WITH THE BIG FOUR

Electricity, Heat, Transportation, and Food

Keeping to a 4-ton annual carbon budget (or whatever goal you set) is going to require that you tackle those daily items that add up big at the end of the year—electricity, home heating, getting to work, and food. Some solutions are simple and won't cost much, at least in the short run. Other solutions involve personal choices and trade-offs, or investing money in home improvements that will pay off in the future. A good target for these big-ticket items is to zero out your electricity footprint and aim for a 1-ton budget each for home heating, work-related commuting, travel, and food.

Let's take a look at some of the daily big-ticket items and strategies to reduce them. This book focuses on things you can do right now that will bring your current direct footprint into line with global climate necessities. Following a "Mediterranean diet" approach to your carbon footprint, you should try to eliminate the high-fat items from your daily diet but try to leave room to enjoy these things as special, life-enriching treats. The last chapter made clear that long-distance travel has to be one of those rare high-carbon treats in a sustainable carbon diet. We'll look at travel separately. And our food diet, it turns out, is a big part of our carbon footprint as well. But running our homes (heat and electricity) and getting to work are the biggest carbon footprint items for most people. To bring our individual emissions down to a defensible level, we need to address those items first.

This chapter will look at how we can get these big-ticket items under control in the short term, without major expense or personal dislocation. My goal is to get those daily carbon expenses as close to zero as possible so that there is room in my carbon budget for life-enhancing luxuries like travel. A zero-carbon goal for life's necessities is good practice, too, because we will need to live on a zero net carbon budget sooner than we think. But we should also consider how these short-term zero-carbon strategies will fare in the longer term, especially if they are adopted by large numbers of like-minded people seeking to reduce their footprint voluntarily or because of government policies (such as carbon fees) that encourage lower carbon emissions. Some actions that are easy to take now may not work in the long term as our entire economy moves toward sustainability.

THE HOME FRONT: ELECTRICITY

Carbon reduction starts with the home. It is often the biggest and most intractable part of the carbon footprint. Our own home in suburban metropolitan New York could be a particularly tough problem, since it seems to epitomize the energy-hogging and carbon-spewing excesses of suburban sprawl. Our home sits on a full-acre lot in suburban Rockland County, an area that is poorly served by public transportation; there is no direct one-seat rail service to Manhattan, for example. There is a convenience store about a half-mile away, but there are no sidewalks most of the way there. The closest supermarket is closer to a mile away. No one in our neighborhood lives without a car. For a long time, we were the neighborhood eccentrics because we had only one car in the driveway. Our house was built in 1937 and has a gas hot water heating system. When we opened up the walls in the living room to rewire the house in 2015, we discovered that there was no insulation—none!—between the stone masonry outer walls and the ancient interior walls.

So, if I can figure out how to reduce my footprint in my mid-twentieth-century sprawl house, anyone can. But our house has one small advantage—its size. With just 1,200 square feet to light and heat, we can live with some built-in inefficiencies. Despite the acre lot, no one would describe our subdivision as tract-style sprawl. In fact, the neighborhood has an interesting history. It was founded as a sort of depression-era exurban commune by socialist economist Ralph Borsodi. Mr. Borsodi bought an old farm and created twelve home sites on land that would be held in common by the community. At the time, the now-defunct West Shore Railroad had passenger service to the Hoboken ferry from a stop a half-mile away. An old flyer advertising the community has the headline "Live in the Country, Work in the City." The community attracted a handful of urban back-to-the land types, who established small, owner-built homesteads. Eventually, the commune dissolved and distributed the land to make a more conventional subdivision: there are sixty houses in the community now. We have one of the twelve original houses; hence its modest depression-era size and construction style. As an added bonus, the 1937 farm fields morphed into mature woods.

Where one lives is one of the most important choices one makes in life. We moved to Rockland County in 1984 and have lived in our current house for thirty years. We raised both our children in this house, and they each had their own room. I didn't move to the suburbs to escape urban ills, or for schools for my children, but just because I wanted to be a little closer to nature than city living allows—a place where it gets dark and quiet at night and wildlife is not far away. We have deer, fox, and turkeys in our backyard, and robins, white-throated sparrows, and wood thrush calls mark the evolution of each spring. Many argue that urbanism must be the solution to climate change—and I can't completely disagree—but the suburbs aren't going away. And at least for now, owning your own home opens up options for footprint reduction that apartment and condominium dwellers don't have, like heating fuel choice and rooftop solar installations. And in my case at least, my small house footprint makes a small carbon footprint easier to accomplish.

When I plug my zip code into an online carbon calculator like https://coolclimate.berkeley.edu/calculator, household utilities for a typical family of four add up to a whopping 14 tons per year: 9.5 tons for electricity and 4.5 tons for heating fuels. Even after dividing this total four ways, 3.5 tons per person for a household footprint does not leave much room in a 4-ton carbon budget for food-related emissions or getting to work. Tackling the household carbon monster in the basement needs to be the first order of business, then.

Electricity Basics

Electricity is typically the biggest part of the home carbon footprint. Thanks to the availability of "green electricity" or renewable energy contracts (see next section), it is also the easiest part of your carbon footprint to eliminate, at least in the short run. Home solar panels also have the potential to eliminate your electricity carbon footprint. Both of these solutions present some surprising carbon accounting issues, and both work well as short-term carbon footprint solutions that might not be a complete solution in the long term.

First, let's get some electricity basics out of the way. When we talk about electricity, we can quantify electrical power and energy with units called watts and watt-hours. A watt is a measure of power, or energy flow, while a watt-hour is a measure of the total energy used: the amount of power flow over time. For those of us who remember those incandescent bulbs we used before we switched to CFLs and then LEDs, it is obvious that a 100-watt bulb is a lot more powerful than a 50-watt bulb. Keeping a 100-watt bulb lit for 30 minutes uses the same amount of energy as keeping a 50-watt bulb lit for a whole hour—the equivalent of 50 watts for an hour, or 50 watt-hours. For household electrical use, utilities bill by the thousand watt-hours, which is called a kilowatt-hour, or kWh. Utilities purchase electricity to resell by the millions (megawatt-hours, or mWh) or billions (gigawatt-hours, or gWh) of watt-hours.

Renewable Energy Contracts

Renewable energy contracts for residential electricity customers are a relatively recent innovation. For most of its history, residential electricity service provided no choice of electricity source to customers. Electric utilities were considered a "natural monopoly" because it never made economic sense for more than one utility to run a duplicate electrical distribution network, complete with a second set of electrical wires down each street, so that consumers might choose from competing service providers. Electrical utilities depend on government sponsorship to get the rights to run service lines in public streets and to condemn right-of-ways for transmission lines. In exchange, state governments heavily regulate the prices charged by electric utilities so that they can't use their service monopoly to gouge customers. Many utilities were organized as local or regional state-owned authorities, like the Long Island Power Authority in New York, the Tennessee Valley Authority in the Southeast, or the hydroelectric "Planned Utility Districts" in Washington State. Far more utilities are private companies, with their guaranteed rate of return providing a steady income (but with little upside potential) for risk-averse investors.

Until relatively recently, electricity generation and distribution were considered inseparable parts of this natural utility monopoly. A typical local electric utility built and owned the electrical generation plants as well as the wires carrying the electricity to the residential consumers. But with the advent of regional transmission networks and wholesale electricity markets for buying and selling among utilities, there was no reason that electricity generation needed to be included in the regulated natural monopoly of electricity distribution. There was no reason that electricity generators could not (or should not) compete with each other. By the 1990s, federal and state utility regulators were allowing independent generators to compete in wholesale markets and were encouraging distribution utilities to get out of the generation business entirely. Once wholesale markets were deregulated and opened to competition, many residential retail markets followed. After all, if the utility could shop for the best generation on regional electric markets, why shouldn't retail electricity customers also be able to shop for generators connected by wire to their local utility? Utilities could become delivery services only, while the product being delivered—electrical energy—could be an independent consumer market transaction between buyer and seller.

These "retail choice" plans are now available in thirteen states and the District of Columbia, serving 16 million accounts. While most electricity sellers in these markets seek to compete for customers based on price, "green energy" providers have entered these markets offering a variety of carbon-free electricity products. Even in states that do not offer retail choice, many local utilities offer a green or renewable energy option to their customers. Green energy costs just pennies more for the retail consumer than run-of-the-grid energy; I am paying about 3 cents per kilowatt-hour more for "premium," "locally sourced" wind power from Green Mountain Energy than I would for the least expensive energy provider available. This would work out to about $18 extra per month for the average customer's 600-kilowatt-hour-per-month utility bill in my area, or about $200 per year; even with my electric vehicles and my electric space heaters, my small house uses well below the average Orange and Rockland Utilities customer. But the lowest-priced energy provider in my market is also offering 100 percent green power that is "nationally sourced," so I could get green energy and save money. Signing up is usually just a matter of going to the green energy provider's website, confirming that they can deliver green energy through your utility, entering your account number, and clicking the "I agree" button.

WHAT'S A REC?

Renewable energy suppliers can be anything from a hydroelectric dam to a massive wind farm on a ridgetop right down to solar panels on an individual homeowner's roof. When a renewable energy supplier delivers electricity to the grid, that energy is more valuable than fossil fuel–generated energy because utilities are required by state regulators to buy a percentage of their electricity from renewable sources and because individuals and companies are willing to pay a little extra to go green. Once the energy is delivered to the grid, it becomes indistinguishable from all the other energy on the grid, so the extra value is represented by a certificate, called a renewable energy certificate, or REC. The REC represents the green energy value of the electricity, but not the electrical energy itself. RECs for nonutility buyers can be traded across regions, so there is no guarantee that a REC you purchase for your home electricity consumption represents electricity that was actually delivered to your grid, unless you can identify the source of the REC.

What if you don't live in a state that allows you to choose your electricity source? Many utilities still offer "green" power options anyway, so you may still be able to purchase green power. And national green energy brokers will offer to sell you green power by purchasing renewable energy certificates (RECs) on your behalf from electricity markets around the country. These RECs are currently very inexpensive. The price for RECs reported by the National Renewable Energy Laboratory for 2016 was just fractions of a penny for a kilowatt-hour (0.035 cent)—working out to about 20 cents per month for the typical electrical bill.

If you can zero out your electricity carbon footprint just by clicking a button on an internet page, why hasn't everyone done it yet? The National Renewable Energy Laboratory reports that 6.3 million electricity consumers purchased renewable energy in one form or another in 2016, including commercial as well as residential customers. This represents a very small fraction of the 126 million households in the United States, and not anywhere near the 43 percent of Americans who worry a great deal about climate change. If you worry a great deal about climate change, why not eliminate the biggest part of your own contribution by clicking a button and paying at most a few dollars more per month on your electric bill? Environmentally aware consumers are willing to pay a substantial premium for organic produce, not just for the health benefits of avoiding pesticide residues, but also for the peace of mind that comes with knowing that organic farming has less adverse impacts on soil, air, and water. Why shouldn't electricity consumers pay a little extra for that same peace of mind?

One reason is inertia. Another is lack of information about green energy options. But another factor is that many greenies pooh-pooh the climate benefits of signing up for green power. You will hear some environmentalists claim that the electrons in your wires are still coming from the nearest, dirtiest coal-fired power plant no matter whom you are paying for electricity. Others will tell you that, no matter whom you are buying your energy from, when you switch on your lights or your air conditioning, your regional electricity dispatcher has to switch on the dirtiest "peaker" energy plant, so your added electricity demand is adding to carbon emissions even if your electricity provider is "green." For these ecopurists, the only true zero-carbon electricity is an off-the-grid house with solar and wind power and battery storage. A general antipathy to utility grid-level solutions and corporate energy markets is also at work. Then the imperfections of our economic system, combined with the impracticality for most of us to go off the grid, become an excuse to do nothing.

While we will take a look at the home solar solution, these ecopurists overstate their case against renewable energy contracts. For one thing, no actual electrons travel from a generating plant to your house. You may remember from high school physics that household electricity is alternating current or "AC." This means that the actual physical electrons in your wires are traveling back and forth between your house and the nearest distribution transformer at the speed of light. That distribution transformer transfers electromagnetic energy from the grid to your local wires, but no actual electrons travel from the grid wires to your wires. And your household load when you switch on your lights or your air conditioner will never be sizable enough to make the utility switch on another fossil fuel–fired generator just for you. In fact, based on the temperature and the time of day, the utility dispatcher is already planning for you and thousands of people just like you to turn your lights on at the exact same time.

Online carbon footprint calculators let you take credit for zero-carbon renewable energy contracts. When you hear about a business taking credit for zeroing out the carbon emissions of its business, very often it bases that claim partly on entering into renewable energy contracts. The fact is, you are buying energy from your electric utility, not electrons. You collect your energy from a vast interconnected grid of regional transmission lines, with all sorts of generators and all sorts of electricity users connected to the same wires. You can withdraw a given number of watts of renewable energy from that network as long as somewhere else in the system someone is putting renewable energy back in. But there is some time-shifting going on as well: it is unlikely that your renewable energy source is adding those watts of energy at the same moment as you are using them. The way these energy markets work, the renewable energy might be put into the system weeks or months away from when you use it. In essence, the grid is also acting as a banking system: a renewable energy generator deposits some energy into the bank, and you make a withdrawal somewhere else.

Our financial banking system works on the same principle: my employer (Pace University) "deposits" my paycheck—represented by some nonphysical computer bits—to my bank account. A week later I can go to any ATM in the country and withdraw my pay in physical twenties.

When I make this withdrawal, I don't worry whether these twenties were used to pay for illicit drugs or even gasoline somewhere in their prior life; it's not "dirty" money, just my money on the network. Renewable energy contracts work on the same principle. If your renewable energy is "nationally sourced," the renewable energy may not even be put into the same regional grid you are withdrawing energy from. So, the grid acts as a sort of energy bank for these renewable energy contracts. Renewable energy seems like a no-brainer for any individuals who care about their climate impacts—switching might even save you money.

What if everyone who cares about climate demanded renewable energy contracts, or at least RECs, in order to zero out his or her electrical carbon footprint? Well, certainly, the competitive price for renewable energy would increase; right now there is a real excess of RECs on the market, which leads to a market price of just a fraction of a cent per kilowatt-hour. An increase in the market price, while bad for renewable energy consumers, would be good for mitigating climate change, since the bigger the premium for renewable energy, the greater the market incentive to build more renewable energy. According to one recent economic study, individual renewable energy contracts are not a sufficiently reliable part of the market to have a significant effect on the number of renewable energy projects. This might change if more consumers committed to such projects. Consumers are willing to pay a 60 percent premium for organic salad mix or a 100 percent premium for organic skim milk. If they paid the same premium for climate-friendly electricity, that would change the incentives.

If a large percentage of consumers insisted on 100 percent renewable electricity, the ratio of variable solar and wind energy might eventually overwhelm the capacity of the existing grid to buffer the imbalance between peak electricity demand and renewable energy supplies. The grid's virtual battery function can only work as long as this time-shifting renewable energy is a relatively small part of the grid. We will discuss this challenge in the "Getting to Zero" chapter. But with renewable energy penetration currently at just about 15 percent of total electricity production (including hydro, which does not have a peaking problem), we are a long way from the 80 percent renewable energy rate where grid adjustment to renewable power becomes highly problematic.

Renewable energy contracts are almost too good to be true—a simple measure to reduce one of the largest elements of your carbon footprint without any lifestyle changes and at a minimal cost. Signing up should be the first step toward getting your individual footprint under control.

NOT ALL GREEN ENERGY PROVIDERS WORK THE SAME WAY

The good news for climate is that interest in green power contracts is on the rise. According to the October 2017 National Renewable Energy Laboratory report, voluntary renewable energy purchases increased by 18 percent to 95 megawatt-hours in 2016. This same report describes a dizzying array of green power options for residential electricity consumers, including utility green pricing, competitive energy suppliers, community choice aggregation, community solar, and unbundled renewable energy certificates, or RECs. Options vary widely among states and between utilities within states. All but the last of these show up on your utility bill if you choose to participate.

With *utility green pricing*, your local utility continues to sell you electricity and delivery services, and it agrees to sell you green energy. That could mean that the utility owns the solar panels or wind farms that deliver power to your regional grid, but it is more likely that the utility buys the power from a third party, or buys RECs representing renewable energy somewhere else in the country.

Competitive energy suppliers and *community choice aggregation* work in a similar way to utility green pricing, except that a third-party energy provider sells the energy and the RECs, which may or not be from the regional grid. Competitive energy suppliers are opt-in: the customer has to choose them. Community choice aggregation is opt-out: consumers are automatically enrolled to purchase green power unless they choose not to. You might already be buying green power without even knowing it!

With *community solar*, a community builds a solar generating facility and delivers the energy to the grid. Subscribers get a credit on their utility bill for their share of the energy delivered.

Unbundled RECs are RECs that are sold separately from the electricity delivered to your house: they are not "bundled" with your electricity provider. Texas, known for its oil and gas industry, also has abundant wind energy and is the nation's biggest seller of RECs.

None of these options automatically ensures that the green energy you are paying for is actually delivered to the regional grid that supplies your house. Any electricity seller can buy fossil fuel energy from a local generator, bundle it with RECs from Texas wind power, and call it green energy. That's why you need to read the fine print on the green energy offers; choose the local generation option if it is available. If you are like me, you want to make sure that you are choosing the renewable energy that is as close as possible to your local delivery utility. That's the most defensible way to take credit for zero-carbon electricity.

Home Solar

If you want to be sure that you are adding as much renewable energy into the local grid as you are taking out, and you own your own home in a sunny location, then rooftop solar panels may be a cost-effective low-carbon option. It is even possible to install solar panels at no cost, as solar power companies will make a deal by which you let them use your roof for their solar panels, and they sell the power to you. This is known as a power producer agreement, or PPA. Financing is available if you want to own the solar panels yourself, and there are also generous federal and state tax credits available for home solar energy systems. The price for a typical home solar system runs between $15,000 and $35,000, depending on the location and the size of the system. If you own the system, you are responsible for maintenance, as well as for repairs if something breaks, like the inverter that converts the direct current voltage from your solar panels to the alternating current used by your house. Leases are also available, whereby the solar installer owns the panels and leases them to you. The Clean Energy States Alliance and the U.S. Department of Energy have published a useful guide explaining these options.

In most parts of the United States, a home solar installation will generate as much electricity during the year as the house consumes, which would allow you to claim 100 percent renewable, local energy. But that doesn't mean you can cut the wire to the electricity grid quite yet. For one thing, your solar panels won't be generating any electricity at night when you need it for lights. Solar panel power production also drops off substantially during the winter, when days are shorter and the sun is lower in the sky. So nearly all home solar systems have to be connected to the grid to balance the offset between solar generation and residential electricity demand. The grid still supplies your energy at night, in the winter, and on cloudy days when your solar panels can't keep up with your energy needs. And since you are tied to the grid, if there is a power outage, you lose power even if the sun is still shining on those rooftop solar panels.

JUST BECAUSE YOU HAVE SOLAR PANELS ON YOUR ROOF DOESN'T MEAN YOU HAVE SOLAR POWER

Say you signed up with one of those "free solar installation" deals whereby the solar installer puts solar panels on your roof at no charge to you and enters into a deal to sell you electricity at below the going rate. Can you then take credit for 100 percent solar power to your house? Probably not. Part of the solar power company's profit comes from selling the renewable energy from your rooftop at premium rates, while providing you with nonrenewable energy from a grid supplier. The accounting for your rooftop power takes the form of "renewable energy certificates" (RECs), discussed in the previous text box. These local-grid RECs are often more valuable to the local utility than the national RECs because many utilities are required by state regulators to purchase a minimum amount of local renewable energy. So you need to check your solar contract carefully to see who owns the RECs from the solar panels on the roof. In some states, even if you own the solar panels outright, the utility ends up owning the RECs if it buys electricity from you under a net metering arrangement. If you don't own the RECs, you can't count the electricity you are using as zero carbon—since someone else is already taking credit for that zero-carbon power!

When you generate more power than you need, your home feeds that power to the grid. When this happens, you may be able to run your meter backwards; in effect, you are selling electricity back to the utility at the same retail price you pay for it. This is called net metering, and it is required by public utility laws in thirty-eight states. As more houses connect solar panels to the grid, solar displacement (at retail rates) of cheaper wholesale power sources for the utility can wreak havoc with the utility's finances, since rates are based on the assumption that residential customers will buy all their electricity from the utility and not generate it themselves. So some states have imposed grid access fees or other charges to offset the reduced energy charges paid by residential solar customers.

A friend of mine installed solar panels on his roof in suburban New York in 2016. He paid for them up front and kept the solar energy credits for himself. The sticker price for the 6-kilowatt system was $32,000, but after the tax credits, he only paid half that amount. The system provides up to 800 kilowatt-hours per month in the summer and as little as 120 kilowatt-hours for the month of December. The system more than covered his electricity usage on an annual basis until he bought an electric car and started charging it at home. So he became a net buyer of kilowatt-hours again—his annual electricity consumption exceeded the output of his solar panels, and he had to pay the utility for the difference.

Home Energy Storage: Batteries and Ice Machines

If you want to cut the power cord and really go "off the grid," you will need some way of storing all that extra solar energy those solar panels make on sunny summer days and use that energy at night, on cloudy days, and during the winter when days are short and the sun is low in the sky. The good news is that available battery technology will let you run your house at night and for a few cloudy days. There is even a home air conditioning system on the market that makes ice during the sunniest part of the day and then uses the ice for air conditioning in the evening as the sun goes away. The bad news is that the batteries are expensive and aren't big enough to store energy in the summer for use in the winter.

Tesla is the leader at this point in offering a residence-sized battery pack, which it calls a "Powerwall." Other electric car makers, such as

Mercedes-Benz and BMW, are also entering the home battery storage market. These batteries use the same lithium-ion battery technologies developed for electric cars. But they are expensive. According to Tesla's online calculator, my small 1,100-square-foot house would require four 13.5-kilowatt-hour Powerwalls to cover a two-day power outage (or two days with insufficient sunlight), at a cost of about $25,000.

This price is just for the batteries; it would not include the cost of solar panels. And I would still have to account for less solar output in the wintertime; solar output in December in the New York metro area where I live is less than half the solar output in July. For my friend who installed solar panels on his house, December output was only one-fifth of August output. So, to truly go "off the grid," I would have to at least double the size of my solar installation and pay for eight Powerwalls to provide four days of backup—pushing $100,000 for the total cost. A more typically sized 2,000-square-foot house would require approximately twice that investment.

For all that off-grid investment, you really couldn't claim that your power profile was any "greener" than a grid-tied rooftop solar panel system. You would still want to be tied to the grid to soak up all the excess power you generated in the summer, or else you would have to waste all that solar energy just when utility solar demand for air conditioning was at its peak. Trying to store all that summer power for winter use would be ridiculously expensive: just multiply the number of Powerwalls needed for four solar-powerless days by about 20 to get through the winter, and you are talking seven figures. You might come up with a cheaper storage system using automotive-style, lead-acid batteries, but these batteries have a short life and bigger footprint—both physical and environmental—compared to lithium-ion batteries.

Another innovative way to balance solar output is by using thermal storage: by taking advantage of peak solar output in the middle of the day in the summer and peak electrical demand on summer evenings. A company called IceEnergy offers an air conditioning system—called the Ice Bear—that uses excess power in the middle of the day to freeze water and then uses the ice in the evening to cool your house. It claims to reduce peak cooling electricity demands by 95 percent. But you would still need a grid tie for lighting and other electricity needs, so it would not let you cut the utility cord. And it only saves you money on your electric bill if your utility charges you more for peak evening energy demand than it pays you for your excess midday solar energy supply—which is not yet the case. Some utilities offer discount pricing for nighttime electrical demand, but none have yet offered discount midday pricing to account for peak solar production.

Home energy storage technology currently works for backup power when the grid goes down, or for a very small house that is not grid connected, or if you are willing to invest a large sum of money to make a statement about independence from utility grid power. We use solar panels and lead-acid battery storage for our cabin in the mountains, but it is tiny (350 square feet) and has no 120-volt AC appliances, just a 12-volt system to power LED lighting and charge our phones and tablets—and the ventilation fan for the composting toilet. It makes little sense at this point to invest in a solar energy system for a regular house, except for short-term backup power, in which case you could have the one quiet, fresh-aired house in the neighborhood while your neighbors' backup generators kick in.

SLASH YOUR ELECTRICITY NEEDS BY DOWNSIZING YOUR FRIDGE

Fridges, like houses, have been getting bigger and bigger, with built-in energy inefficiency. Buck this trend! After our kids moved out of the house, we ditched our mid-sized fridge and replaced it with a waist-high, hotel-style mini-fridge. Face it, the big fridge was just full of mildewing old leftovers and freezer-burned ice cream we would never eat. Our mini-fridge holds nine cans of beer in a handy dispenser, enough milk and coffee cream to get us through the week, with enough room for occasional meat and poultry treats and cheeses. When we want an ice cream treat in the summer, we take the excuse to bike or walk to the community ice cream shop. Fresh veggies keep fine out of the fridge for a day or two. If we need to cool a lot of beer for a party in the summer, we buy some ice for the cooler. Our new mini-fridge cost just $160, and it only uses 225 kilowatt-hours per year (according to EPA ratings). That's 600 kilowatt-hours per year less than the average new full-sized fridge with an icemaker—equivalent to 1,200 pounds of savings in carbon emissions from average grid electricity in our region. And if you aren't ready to downsize to a smaller refrigerator, just turning off the icemaker in the refrigerator you have in favor of ice cube trays could save you several hundred kilowatt-hours per year.

Efficiency and Reduction

Most carbon footprint discussions start with energy efficiency and reduction as an approach to lowering your individual footprint. I am saving it for the end of the electricity discussion, because if you are reading this book and have got this far, further appliance efficiency and reduction measures are not, by themselves, going to get your electricity footprint down to the range needed for a defensible carbon budget. You have probably already replaced your incandescent bulbs with energy-efficient LED bulbs, reducing your lighting electricity demand by a whopping 75 percent. But lighting is only about 10 to 15 percent of home electricity demand, so those LED bulbs might shave just 200 pounds from an individual's share of a typical 10 tons of annual household electricity carbon emissions in my area. Every little bit counts, but those savings are not going to get your annual carbon footprint under 4 tons. Apart from lighting and air conditioning, your refrigerator is usually the biggest energy hog in the house. Trading in an old refrigerator for the latest EnergyStar model will save your household about 200 pounds of carbon emissions per year, according to the EPA EnergyStar website. Your individual carbon savings will be about 50 pounds per year. Trading in your washer for a new front-loading model will also save electricity and water.

Trading in these appliances is worth doing; it will save you money in the long run, as well as carbon emissions. So is taking a hard look at your energy habits and making a conscious effort to stop wasting energy: use a little less air conditioning (if you have it) and turn off the entertainment system in empty rooms. But all of these measures, taken together, still won't get your electricity footprint below 1 ton in a 4-ton carbon budget. And if you add electric vehicle charging to your electricity demand, that will only increase your electricity footprint if you are still purchasing conventional electricity.

So even with all the efficiency measures you implement, you will still probably want to sign up for solar panels on your roof or a renewable energy contract from your utility in order to meet your carbon budget. This raises an accounting paradox: if you can count a renewable energy contract or rooftop solar as zero-carbon emissions electricity, your electricity-related emissions are still zero no matter how inefficient your appliances are or how much electricity you waste. So renewable energy takes away some of the incentive to reduce energy consumption. But you still want to keep good energy habits; wasted electricity still costs you money, even if it is renewably sourced. And (as we will discuss below) in the long run, as we transition to a zero net emissions fuel economy, energy costs will likely rise significantly, making energy efficiency and conservation an economic necessity. So, turn off the lights when you leave the living room!

THE HOME FRONT: HEAT AND AIR CONDITIONING

In the southern half of the United States, air conditioning in the summer is a bigger part of your energy demand (and carbon emissions) than heating in the winter, and your air conditioning carbon emissions will become part of the electricity-related emissions just discussed. But for those of us in the colder climates in the north, heating-related emissions are a big part of our home carbon footprint. Although many homes have electric heat, gas or oil heat is usually much cheaper at current energy prices—but at a huge carbon emissions price. Unlike your electricity-related emissions, there is no easy fix for your oil- or gas-related carbon emissions: there is no zero-carbon renewable energy option, at least not yet. Cost-effective carbon reductions are possible through efficiency and smart thermostats, but they may not slash heating emissions enough to be consistent with a 4-ton individual carbon budget. You may have to change the way you heat your house.

How Far Can You Get with Efficiency and Lowering Your Thermostat?

Improving your home's efficiency may be a cost-effective way to get significant savings on your carbon footprint and your heating bill. Replacing the windows and doors in your house with new high-R-factor insulated windows and doors should reduce your heating energy demands by about 15 to 20 percent for a typical house. It might cost $8,000 to $20,000 up front, but it will reduce your energy bill. The payback on this investment, however, will take over ten years. Consumer Reports does not recommend replacing your windows to save money on heating and cooling.

You could also improve the thermal efficiency of your house by adding insulation, but this will be more disruptive—and expensive—since you will have to open up the walls of the house to remove the old insulation. To superinsulate the house, you would have to build out the walls to make room for thicker insulation. Unless you do the work yourself, the cost will likely run into the tens of thousands of dollars—and you can expect a 10 to 20 percent reduction in energy use, depending on the climate where you live. This might be worth doing if you are already planning a gut rehab of your house, but if not, this may not be worth it.

How about lowering your thermostat? According to EPA, lowering the temperature in your house by 7°F to 10°F for 8 hours a day (when you are at work or school, for example) can save as much as 10 percent a year on heating and cooling compared with the system's normal settings. You would most likely want a programmable thermostat that can start heating or cooling an hour before you return home so that when you arrive it is a comfortable temperature. By adding the insulation plus a smart thermostat, you can get 18 to 28 percent savings.

There is also something called zoned systems. That is when each room in a home becomes a separate zone for heating and cooling, with its own thermostat. The benefit is that rooms that are used less are not heated or cooled, while rooms that are most occupied are a more comfortable temperature. This is an expensive option that makes the most sense when your current system needs to be replaced. But if not all rooms in your home are heavily used, this can increase the energy efficiency of your home heating and cooling system by around 15 percent, depending on where you live and how large your house is. This is most effective in a very large home with many rooms.

Combine all of these cost-effective measures (insulated windows and doors, comfortably lower temperatures, zone heating), and you can hope to reduce heating energy by as much as 38 percent. At the high end, this would reduce the average 1.5-ton individual heating footprint in my area to 0.93 of a ton—a significant reduction. This might be enough to fit in a 4-ton direct budget, depending on your other carbon emissions. If not, you may need to think about heating your house differently.

Heating Without Oil or Gas

Home heating is one of those individual activities, like driving a gas or diesel car, in which we burn fossil fuels directly. If we really want to go fossil free, the best way to do it is to stop buying and burning the stuff. For home heating, this means exchanging your gas or oil burner for some sort of nonfossil heat: electric heat from renewables, geothermal heat, or wood heat in the form of logs or pellets. Each has short-term advantages and disadvantages, as well as long-term challenges. For this discussion, we are assuming you are looking to retrofit an existing house, not building new, so we will not consider superefficient, passive solar heat exchange systems that can approach zero-energy climate control but cannot feasibly be retrofitted to an existing house.

Convert to Electric Heat

Perhaps the least expensive heating conversion option (at least in up-front costs) is simply to convert your house to electric heat. Electric heat is the cheapest heating system to install; that is one reason that homebuilders build new houses with electric baseboard heat, and electric heat is the most common heating option in the United States. But homeowners with electric heat pay extra in the long run: monthly electric heating bills are far more expensive in dollar terms than the equivalent monthly bills for heating with natural gas, propane, or oil. Part of that cost advantage is artificial: gas or oil heat causes climate costs (externalities) that are not included in the price of the fuel.

If climate impact reduction is your number one priority, and you can purchase renewable electricity, then the extra cost of electric baseboard heat is well worth it. Electric heat is 100 percent efficient: all of the electrical energy used is converted to heat. Expect to pay about $3,000 to $5,000 to install baseboard heat in your house. In late 2017, the U.S. Energy Information Agency projected that the average household would pay $644 annually to heat their home using natural gas and $980 annually for electric heat that winter, a premium of $336 per winter to heat your house with electric heat. You will probably have to bring in an electrician to add the extra wiring to handle the heating load.

Not ready to commit to installing a redundant heating system in your house? You might consider the less expensive alternative of setting the thermostat for your oil or gas heat system low and purchasing a few of those oil-filled electric radiator-style space heaters to keep individual rooms at a comfortable temperature while the fossil heat does the work of keeping the pipes from freezing.

If everyone in the Northeast suddenly switched to electric heat, the increased winter demand might pose problems for the electric grid. While peak electricity demands usually occur during summer air conditioning season, extremely cold winter nights also see high demand. In fact, utilities with a high proportion of electric heating customers, like Houston, experience their annual peak demand during the winter heating season, and not the summer air conditioning season. We'll consider the long-term challenges of conversion to electric heat and all-renewable energy in chapter 10. If a large percentage of homeowners switched to electric heat and insisted on renewable energy contracts, that would swell the demand for renewable energy and the corresponding RECs, resulting in some combination of higher prices for renewable energy (bad for consumers, good for renewable energy generation), increased renewable energy generation to meet this demand, or limited availability of renewable energy options for consumers. And, as discussed a few pages back, your rooftop solar panels will not be able to keep up with your heating demand, not when they are only producing one-quarter to one-half of the solar power they produced in the summer.

Heat Pumps: Geothermal and Air Sourced

True geothermal heat consists of running pipes into steam vents and hot rocks in geologically active zones. This kind of heat is "free" energy: you can tap into a limitless supply of energy buried beneath the earth's surface. This heat energy comes both from radioactive decay of the continents and residual heat deep in the earth left over from the earth's formation as a blob of molten rock. This option is available to very few people in the United States; if you do not live close to a natural hot spring or geyser, your house does not qualify.

But a geothermal heat pump is the next best thing; it manages to beat even the 100 percent efficiency of electric resistance heating systems (like baseboard heat) by taking advantage of the relatively constant temperature of groundwater lying a few dozen yards beneath the ground. This groundwater maintains a constant 50°F or so throughout the year. You might wonder how 50°F groundwater could possibly help keep your house a comfortable 65°F. The answer lies in what experts call a heat pump, which is just a fancy term for the refrigeration unit on every refrigerator and air conditioner. Yes, that heat pump makes the fridge cold, but it does so by making the air around the compressor warm, effectively "pumping" the heat out of your icebox and into the air. Turn this heat pump around, and you can warm up the air in your house by pumping some of the heat out of that 50°F groundwater—making it drop to, say, 49.5°F instead. Because water holds much more heat per unit volume than air, a small amount of heat pumped out of the water turns into a large temperature increase in the air. This allows ground-sourced "geothermal" heat pumps to be as much as four times more efficient than traditional systems like baseboard heat or oil-filled electric radiators. As an added bonus, these heat pumps are even more efficient when you turn them around and use them as air conditioning units in the summer, since that groundwater is closer to the temperature you want your house to be.

Installing a geothermal heat pump is a major home improvement project, as you will have to drill wells into the groundwater or excavate your yard to place a heat exchanger deep in the ground. Also, you cannot retrofit a geothermal heat system to hot water or steam radiator pipes;

it only works with forced hot air heating ducts. So, if your house already has central air conditioning, you are in luck; otherwise, you are looking at a very expensive home renovation to make space for heating ducts throughout the house. Expect to pay somewhere from $20,000 to $25,000 to retrofit an existing central air conditioning system to a geothermal heat pump. You can expect to save 40 to 60 percent on your utility bills (compared to electric HVAC systems).

Electric geothermal is probably not cost-effective yet if you are just seeking to lower your heating bill, but if you have central air, it has the advantage of both saving you money on air conditioning and transitioning away from fossil fuels for heat without increasing your heating cost.

In recent years, air-sourced heat pumps have improved, resulting in greater efficiency and lower temperature limits. These units can heat your house by working like air conditioning systems in reverse: instead of making the outside air hotter to cool your house, they make the outside air colder to heat your house. These units can now approach the efficiency of ground-sourced heat pumps, but without the need for expensive excavation and well construction. What's more, newer "mini-split" heat pump units don't require ducts and can be retrofitted to existing homes that don't already have central air conditioning ducts. These air-sourced heat pumps have a lower temperature limit of about 0°F. In colder climates, installers recommend that you keep a backup system: electric resistance heat, a woodstove, or your existing boiler. The mini-split units are best for smaller and well-insulated houses with open plans that allow natural air circulation. The cost to install a mini-split unit runs from $3,000 to $5,000 per unit. A small house might get away with one unit, while a mid-sized house might require three units. On the horizon are superefficient CO_2-based, air-sourced heat pumps that work in even colder temperatures; there is already one unit marketed by Sanden that uses this technology to run an electric water heater that claims to be six times more efficient than a conventional electric water heater.

Nearby friends of ours converted their house to a geothermal HVAC system and regretted it. They paid about $70,000 for the installation, which involved digging up their entire backyard to install a series of wells. Contractor costs in the New York metro area are much higher than the national average, and their installation cost reflects this fact. Their actual cost was reduced to about $40,000 after they received tax credits and rebates. The system works great for air conditioning in the summer, but they found that in the winter the geothermal heat pump did not work in temperatures below about 20°F, and they found themselves paying $600 per month in electric bills to run the backup electric heating system. Other friends of ours who live in a rural area in upstate New York had a much better experience with a different contractor, and they are saving money while keeping their house comfortable. Given the advances in air-sourced heat pump technology, ground-sourced heat pumps probably only make sense for larger homes in colder climates and where excavation costs are low.

Wood Heat: Outdoor Wood Boilers and Woodstoves

Wood home heat can be produced by a wood boiler connected to a hot water heating system, or a woodstove. Either may be fired by cord wood or by wood pellets. Wood-fired boilers can be located indoors or outdoors. Because they are easier to install, outdoor wood boilers are popular in rural areas. I have saved wood heat for last in this discussion because it may be a tenuous carbon reduction strategy. Wood heat is controversial among environmentalists, so you should approach conversion to wood heat with caution. Because of its moisture content, wood heat actually emits more carbon dioxide per BTU than either gas heat or oil heat. Some of the energy has to go for boiling off the water in the wood. So the low-carbon attributes of wood heat depend entirely on offsetting the carbon emissions with simultaneous carbon sequestration by the trees and forests that produce the wood. Wood heat can be considered carbon neutral only so long as the wood is being produced sustainably.

WHEN A TREE FALLS IN THE WOODS, IS IT BETTER TO BURN IT OR LET IT ROT?

Wood heat can only be considered carbon neutral if the forest it comes from is storing carbon faster than the trees are being harvested. But what about deadwood? Dead trees left to rot will decompose—converting the cellulosic carbon in the wood to both CO_2 and methane, a greenhouse gas that is 20 times more potent. But some of that carbon in the deadwood would also be locked up in soil carbon: a valuable carbon sink that reduces atmospheric greenhouse gases. The exact balance between methane and CO_2 releases and stored soil carbon is hard to determine. But harvesting deadwood will always achieve a better carbon balance than cutting live trees, since at least some portion of the carbon in deadwood is destined to add to greenhouse gas emissions whether your burn it or let it rot.

The climate sustainability of wood heat thus comes full circle with Gifford Pinchot's utilitarian quest over a century ago to introduce sustainable forestry to the United States. And that's the rub for environmentalists: a monoculture forest managed for maximum sustainable wood yield is not a natural forest, so every acre of land dedicated to wood production is an acre lost to the diverse natural habitat needed by wildlife. Forest diversity is also necessary for the resilience needed if natural forests are to have some chance of surviving and adapting to climate change. When the European Union recently declared that wood-generated electricity would be considered carbon neutral for the purposes of its carbon emissions limits, timber interests in the southeastern United States introduced plans to build plants to pelletize market timber and ship it to Europe as fuel. Local and regional conservation groups have protested, pointing out the threat to natural forests, the lack of assurance that carbon emissions will be sequestered, and the failure to account for the carbon emissions from oil-fueled pellet ships.

Nevertheless, wood heat might be defensible as a carbon-neutral option for heating your home, depending entirely on where your wood is coming from. I think wood heat is at least arguably carbon neutral under the following circumstances: if your wood source does not cause cutting live trees or if your wood source is based on sustainable natural forest management.

Ideally, your source for wood would not involve cutting any live trees. This might be the case if you have a woodlot near your house where you can harvest dead trees and storm blowdown in sufficient quantities. But if that is not possible for you, transporting wood long distances by fossil-fueled transport quickly adds to its cost—and its net carbon emissions. So the woodlot option works best in rural areas, though in our suburban county the Palisades Park Commission grants permits to remove the plentiful deadwood in heavily wooded Tallman State Park. In suburban areas, landscapers and arborists sell cordwood derived from trees they are hired to cut and clear for landowners and parks. This wood might also be considered a carbon-free windfall, since your demand for cordwood doesn't directly cause the storm blowdown or homeowners' removal of their trees. (It might be argued that an arborist who can command a high price for cordwood will charge less for tree removal, adding to the number of homeowners willing to pay to have trees removed.) And your wood source might also fall in this category if it consists of pelletized wood scraps from lumber mills that would otherwise be burned or landfilled. Most commercially marketed wood pellets for home stoves claim to be produced from this sort of "anyway" scrap. Those trees were going to be felled for lumber or pulp anyway, so the wood pellets are a windfall, and burning them in your home stove doesn't add to emissions.

WHAT CAN APARTMENT DWELLERS DO TO CUT THEIR HOME FOOTPRINT?

The good news for apartment dwellers is that multi-unit residential buildings are more efficient to heat and cool per unit of floor area because the heat that leaks from one unit's ceilings helps heat the apartment one flight up instead of radiating to the stars through the roof. One estimate suggests that multi-unit apartment buildings should be twice as efficient as a single-family house. But apartment dwellers have little or no control over their heating carbon emissions; sure, you can turn your thermostat down, but if your building has an oil-fired boiler in the utility basement, you are stuck with 2.5 tons or so of carbon emissions per person. Your footprint-reducing efforts will have to be directed to using other efficiency measures and convincing the landlord or co-op board to convert to a more climate-friendly heating system. If you live in a city apartment, at least you may be able to minimize your commuting footprint by biking or using mass transit.

If your wood source involves cutting live trees, then it is only carbon neutral if you have an assurance that the rate of cutting does not exceed the long-term regrowth of the trees. The least environmentally harmful way to ensure this would be to selectively cut dead and damaged trees in a diverse natural forest. This kind of harvest of a woodlot might require 5 to 10 acres of woodlot to heat a typical house for the winter. Once again, if you live in a rural area and can cut wood from your own woodlot without traveling too far, this might work, or if you can buy cordwood from a nearby woodlot managed this way.

Few commercial forests are managed this way, as this approach does not lead to the maximum number of board feet (or cords of wood) in annual yield. Commercial foresters prefer to clear-cut an area and plant one fast-growing species of tree (a "monoculture") to replace the clear-cut forest. If forests were managed for maximum BTUs of wood heat potential per year, fast-growing species such as black walnut and black cherry would be the species of choice, not the native trees such as hickory and oak in the Southeast or maple, oak, and beech in the Northeast. Forests that are harvested by clear-cutting take many decades to re-sequester the carbon emitted by burning the wood that is harvested; in fact, a clear-cut forest is a net emitter of greenhouse gases for five years after harvest because of the carbon and methane emissions from the branches and treetops that are left to rot.

Depending on how well managed the forest source is, you might be able to claim a carbon emissions advantage from trading in your gas or oil heat, but it may come at a great environmental cost in habitat diversity. Every low-carbon energy option involves some sort of environmental trade-off: wind turbines kill birds and destroy the wilderness aesthetic of ridgelines and open ocean, while mining for the metals built into batteries and solar panels destroys the landscape and releases toxins. And fossil fuels themselves cause vast ecological harms in addition to the climate impacts of burning them: poisoned groundwater results from hydrofracking natural gas, coal mining converts vast mountaintop landscapes into rubble, and oil production poisons land as well as oceans. There is no way objectively to compare the ecological harms of using fossil fuel alternatives to the ecological harms of continuing to use fossil fuel. So you will have to make your own choice about whether your wood source constitutes an acceptable environmental trade-off for climate change mitigation.

Wood pellet stoves or boilers may be a sustainable alternative as well, at least for the time being. Most retail wood pellet manufacturers claim that their pellets come from lumber mill scraps that would otherwise be burned or landfilled; if this is true, pelletizing the wood scraps and burning them for heat adds no net carbon emissions. But if wood pellet demand increases enough that suppliers start using whole trees, wood pellet heat would have the same carbon sustainability issues as cordwood.

Wood heat involves other trade-offs as well. Except for pellet stoves, wood heat requires regular attention: starting the fire in the stove or the boiler and constantly stoking the burner with new wood. You still need a backup heating system for when you are not home to feed the stove. Wood heat also emits air pollutants other than greenhouse gases; if everyone in a populated area switched to wood heat, the air would violate air quality standards. Wood-burning stoves and fireplaces have been banned by many municipalities across the country for this reason. Wood burners also emit heavy smoke plumes when they are started cold—outdoor wood boilers are particularly offensive—and your neighbors will not appreciate it. But if you live in a less populated area, or a populated area where few of your neighbors use wood heat, and you have a sustainable source of wood fuel, a glass-fronted woodstove can be an aesthetic pleasure that also reduces your carbon footprint.

Cooking Without Gas and Replacing Other Gas Appliances

I have focused on reducing your heating footprint because for those of us in northern regions with cold winters, heating is the biggest source of home fossil fuel consumption and corresponding greenhouse gas emissions. But you may burn natural gas or propane for other home appliances, particularly your clothes drier, kitchen range, and water heater. Compared to home heating, these appliances have a relatively small carbon footprint, so if you can get a grip on your heating carbon tab, you can probably meet your carbon budget even without replacing your water heater, range, and drier with their electric counterparts. In addition to the direct replacement cost, expect to pay more on a monthly basis for electric bills than you will save on gas or propane: electric appliances are currently more expensive to operate than equivalent gas appliances.

You might also consider renewable options for clothes drying and water heating. The wind-powered solar clothes-drying system has been around as long as clothes have, and it is dirt cheap—it's called a clothesline. You pay with a little bit of extra time hanging clothes up (but since you can sort your clothes at the same time, it's not all lost time), and outdoor clothes drying doesn't work on rainy or cold days, so you will probably need a backup dryer or an indoor space for drying clothes.

Solar hot water systems are simple and efficient; rooftop solar water heaters operate at a much higher efficiency rate than solar electric panels. In the summer, a solar water heater will make your water plenty hot for domestic use even without a backup water heater. Even in the winter in northern climates, solar water heat will reduce your energy consumption. But installing the system may be a major construction project: your roof may not be strong enough to store the 100 gallons or so of water in the solar heating system. Expect to pay between $8,000 and $10,000, including installation, for the typical solar water heating system.

What Worked for Us

I got serious about reducing my personal carbon footprint in 2007, and I started by looking into our household carbon footprint. I called a solar panel installer to look into rooftop solar panels. The solar rep looked at the mature walnut and ash trees shading our house and shook his head; I never heard back from him with an estimate. Instead, I looked into reducing our heating bill, and I spoke to our neighborhood heating system installer about updating our 1937 boiler with a modern, efficient unit. The new unit claimed an annual fuel use efficiency (AFUE) of 80 percent—a 30 percent improvement over the 60 percent AFUE rating of old boilers (95 percent AFUE units are now available). But I can't say I noticed much savings in our gas usage or heating bills the following winter. We had already minimized our heating bills with a programmable thermostat that set the temperature low overnight and during the day when the house was empty.

I started keeping track of my carbon footprint on a regular basis in 2014. I took another look at our house footprint. It was easy, by then, to sign up for a 100 percent renewable electricity contract and zero out our electricity carbon footprint. I also revisited home heating options. Our old house has a huge brick fireplace, and we enjoyed nightly evening fires each winter. There was always plenty of downed wood from our backyard. The fireplace in our house employed cutting-edge 1930s energy technology: "heatilators" that vented outside air into the room through the steel fireplace back, prewarming the outside air needed to feed the chimney draft.

But I knew that open fireplaces were inefficient, venting nearly as much cold air into the house as they produced in heat. So, it was time to look into a woodstove insert. I found one with a glass front that would fit flush into the existing hearth, so we continue to enjoy the aesthetic glow of burning wood collected from our backyard. An air blower comes on when the stove gets hot enough; the manufacturer (Napoleon Stoves) claims 55,000 BTUs of heat output, suitable for heating 1,500 square feet. It keeps our living room warm enough that in a typical winter the gas heat no longer comes on as long as we are home to feed the stove in the morning and the evening. Since our bedroom is directly over our living room, enough heat rises to warm that space as well. We use an oil-filled electric space heater to warm up the other spare bedrooms when someone will be using them.

The stove itself cost us about $1,000 delivered. I saved money by installing the stove and the needed flue liner myself (this involved drawing up plans and getting a permit and inspection from the local building inspector). It looks like the woodstove insert has cut my heating bill by about $200 each winter, so the woodstove has already paid for itself financially as well as having reduced my home heating footprint down to about 100 pounds of CO_2e per month. As this book goes to press, I have arranged to install an air-sourced, mini-split, two-zone heat pump to supplement and automate the woodstove. This will cost me about $10,000—which will pay for itself in carbon footprint reductions, but not so much in dollars.

For a summary of home heating options, see tables 8.1 and 8.2.

TABLE 8.1 Heating Options for 1,800-Square-Foot Home with Natural Gas Heat (National Averages)

	INSULATED WINDOWS AND DOORS	LOWER THE THERMOSTAT	ELECTRIC BASEBOARD (WITH RENEWABLE ENERGY CONTRACT)	ELECTRIC HEAT PUMP (WITH RENEWABLE ENERGY CONTRACT)	WOOD (WITH SUSTAINABLE SOURCE)
Up-front cost	$3,000–$7,000 for ten windows in single-story house	None, other than warmer clothes	$3,000–$5,000	Air sourced: $5,000–$20,000; Ground sourced: $20,000–$25,000	$3,000–$4,200, including stovepipe and installation
Annual heating cost change*	$100 less	$30–$100 less	$340 more	$50–$250 less	$850–$1,350 more
Percentage energy use reduction	Up to 15% lower	5%–15% lower on your heating bill by setting thermostat 10–15 degrees cooler for 8 hours a day	No change	40%–60% lower	No change
Tons of CO_2 reduced	0.51	0.17–0.51	3.4	3.4	3.4

*Will vary by region.

TABLE 8.2 Heating Options for 1,800-Square-Foot Home with Oil Heat (National Averages)

	INSULATED WINDOWS AND DOORS	LOWER THE THERMOSTAT	ELECTRIC BASEBOARD (WITH RENEWABLE ENERGY CONTRACT)	ELECTRIC HEAT PUMP (WITH RENEWABLE ENERGY CONTRACT)	WOOD (WITH SUSTAINABLE SOURCE)
Up-front cost	$3,000–$7,000 for ten windows in single-story house	None, other than warmer clothes	$3,000–$5,000	Air sourced: $5,000–$20,00; Ground sourced: $20,000–$25,000	$3,000–$4,200, including stovepipe and installation
Annual heating cost change*	Up to $220 less	$70–$220 less	$500 less	$900–$1100 less	$50–$550 more
Percentage energy use reduction	Up to 15%	5%–15% lower on your heating bill by setting thermostat 10–15 degrees cooler for 8 hours a day	No change	40%–60% lower	No change
Tons of CO_2 reduced	1.73	0.58–1.73	11.5	11.5	11.5

*Will vary by region.

GETTING TO WORK: THE ZERO-CARBON OPTIONS

After heating, lighting, and cooling your home, getting to work and back is likely the largest part of your routine carbon footprint. The average American commutes 8 to 10 miles each way to work, and that gallon a day of gasoline for the typical solo commuter in the typical car adds up in carbon emissions over the year. And as we explored in chapter 7, even if you are taking a train or bus most of the way to work, there are emissions associated with public transportation as well. Low- and zero-carbon commute options range from trading your fossil-powered vehicle in for an electric vehicle to ditching your car commute for a bike or public transportation.

Electric Vehicles (EVs)

When I started my carbon budget project, the only electric car on the market was the Tesla Roadster. I test-drove one when Tesla had a demo day at our boat club. It has heart-pounding acceleration, every electronic gizmo built in, and a truly luxurious finish. But with a sticker price north of 80 grand, it was just not an option for me. But now most of the major car companies have electric vehicles in their lineup, with prices that compare favorably to the average new passenger vehicle price. For example, the 2018 Chevy Bolt has a sticker price of $37,500, but with the federal tax rebate, the actual cost is a little over $30,000—well below the average new vehicle price of $36,000. With a 238-mile range on a charge, even those with "extreme" commutes can make it to work and back on one charge. Even the most limited-range EVs (like the little Smart-for-Two I drive) have ample range for the typical commute. And Tesla is now delivering its Model 3 cars at a price more in line with the typical passenger sedan.

Want heart-pounding acceleration at a compact car price? You might consider an electric motorcycle (e-moto). Zero Motorcycles in Santa Cruz, California, offers electric motorcycles priced around $14,000, with ranges up to 200 miles on a charge. The energy efficiency of an e-moto is hard to beat: the equivalent of up to 300 miles per gallon, compared to about 100 miles per gallon for an electric car. Harley Davidson is releasing its own e-moto in 2019, with a sticker price close to $30,000. But unless you live in a sunny climate, there will be days when you can't ride a motorcycle to work. And you can't take the soccer team to practice on an e-moto (your kid's teammates' parents will forbid it). So you may need a backup plan for your e-moto commute.

Combined with some form of renewable home electricity source, EVs count as a zero-carbon commute option. But don't count on your rooftop solar unit powering both your electric vehicle and your household electricity needs, at least not in the winter months. The 5 kilowatt-hours or so per day your EV will need (for a typical commute) will add substantially to your household electricity demand.

Paddle, Walk, or Bike

The other zero-carbon commute option is good old human-powered transportation. Some advocates will tell you that getting to work under your own power is the only climate-responsible way to commute. As someone who gets to work under my own power on a regular basis, I am not ready to go that far. There can be real challenges for pedal-powered commuters; sheer distance and time involved are not the least of them. But a 2018 study found that non-bicyclists vastly overestimated the time it would take to bike to work. In my experience, 13 miles an hour is an average time for a reasonably fit non-athlete. So that typical 10-mile commute to work will take most of an hour, while even the most congested 10 miles in traffic won't take much more than a half-hour. Census Bureau statistics peg the median automobile commute time at 26 minutes.

Let's get the negatives about bicycle commuting out of the way. It is physical work in a society that shuns physical work; this should be a positive, but it is a psychological barrier for many. You have to be reasonably fit to bike 20 miles a day. In many cities and most suburbs, there is little or no bicycling infrastructure, and that can be dangerous. Bike paths, if they exist, seem to run perpendicular to the direction you need to go, forcing you to share the road with rush hour traffic. For unfathomable reasons, many drivers seem to resent the presence of an adult on a bicycle and will take it out on you by honking their horn next to you, yelling obscenities, or driving with two wheels on the meager shoulder available for you to ride out of traffic. If you "take the lane" on a bike at 15 miles per hour, as you are entitled to do, the wrath of the drivers stuck behind you will be palpable. Bicycling accident statistics on a mile-per-mile basis are hard to come by, but I feel less safe in traffic on my bike than I do on my motorcycle; at least I get some respect from car drivers when I take the lane at traffic speeds on my e-moto. Moreover, if you are a parent with children in preschool, it is not easy to pick them up and drop them off with a bicycle. And if you have children in school, you may not be comfortable being unable to get in a car and get to them quickly in an emergency.

Finally, as with the motorcycle, there will be some days in northern climates when you just can't ride a bike to work. Biking in traffic in the dark is hazardous, especially when you can't see the potholes along the edge of the road where you are forced to ride. There are plenty of die-hards who don the gear and bike to work in rain, sleet, snow, and dark of night, but it's not going to appeal to many. At the opposite extreme, in hot weather you will need a way to freshen up when you get to work so you don't smell like a Tour de France team. So most of us will still need a backup commute plan even if we make biking a regular low-carbon commute option.

NO-SWEAT CYCLE-COMMUTING WITH E-BIKES

You say you'd like to bike to work, but you aren't ready for physical training? No sweat. You can get an e-bike. Combining advances in lithium-ion battery technology with old-fashioned cranks and chains, an e-bike adds an electric motor assist to your pedal cruiser. The effect is a little unnerving: light pressure on the pedals gives you almost superhuman acceleration and flattens out the hills effortlessly. U.S. e-bikes are programmed to max out at 20 miles per hour, but that is still a bit faster than the average person pedals the average bike. And e-bikes sip energy; by some calculations the carbon emissions of generating the electricity used by an e-bike is less, mile per mile, than the food-related carbon emissions needed to supply the calories burned by a human-powered bike. State regulation of e-bikes varies, so make sure they are legal in your state before hitting the road.

And for all that, biking to work is not necessarily a zero-carbon option. If you make up the extra calories you burn with a typical diet, your food-related emissions might be the equivalent to a 120-miles-per-gallon car (based on the typical American diet). Since most Americans consume more calories than they need to maintain their weight, you don't need to make up all the calories you burn bicycling, but your food intake (and food impacts) will go up.

But all of the negatives of bicycling have corresponding positives. If you are not in shape, biking to work will get you in shape in a relatively painless way (just bike slower to start out). Your increased fitness will make you feel better about yourself and will make you a happier person. If you already dedicate a half-hour per day to exercise, the time you spend biking to work instead will not be subtracted from family or leisure time. Many cities and suburbs (notably including New York City) have improved their bike-friendliness with protected bike lanes and bikeways. Like other forms of moderate outdoor exercise, bicycling promotes a zen-like meditative state that reduces stress throughout the day and improves cognition. (Yes, bicycling out of doors really does make you smarter.) Bright rechargeable lights are now available to light the road ahead and make you obvious to motorists in the dark. And resourceful cyclists have figured out how to freshen up at work even without a shower, using Combat Wipes and dry shampoo.

If you live closer to your place of work, walking can offer all the stress-relieving benefits of outdoor exercise without the traffic hazards. But walking isn't practical if your commute is more than a mile or two; 3 miles in an hour is a pretty brisk walking pace.

I can't possibly walk the 15 miles from my home to my office. For me, the big challenges for biking to work are the weather and the Hudson River. The old Tappan Zee Bridge has no bike lanes, and the nearest bridges that allow bikes would require a 40-mile detour on my 15-mile commute. My solution for the past twenty years has been to paddle a kayak 3 miles across the Hudson and bike the remaining 8 miles to White Plains. But if you think a bicycle is slow compared to a car, try paddling a kayak at 4 miles per hour. It takes me 2 hours to get to work this way (compared to a 40-minute drive in traffic). It also requires me to have two bicycles, one on each side of the river. But my full-body workout feels great by the time I get to the office.

Paddling across the Hudson River is not an option in the winter when the landing docks are pulled out and ice chokes the channel. Even in the summer, afternoon thunderstorms are a hazard. Between winter weather, darkness, storms, and the 4 hours it takes for the round trip, my paddle-and-pedal commute is limited to a few days a week in the warm and light half of the year. But the Thruway Authority started a replacement Tappan Zee Bridge project in 2013 that will include a bike lane (not yet open) for the 3-mile crossing. When the new Tappan Zee bike lane opens, I could probably make the bike commute in about an hour, and with a good headlamp, a year-round cycle commute will be possible. But there will still be ice storms.

Public Transportation

Most environmental groups will advise you to ditch your car for public transportation. But if your daily carbon footprint is your primary concern, then switching to an electric vehicle or a bike will make a bigger reduction in your footprint. There is no zero-carbon public transportation option yet. The New York City subway system has just about the lowest carbon emissions, at 0.12 pound per passenger mile traveled, so if you are on a subway line, your commuting footprint will be hard to beat. But commuter rail emits more than twice as much per mile, and city buses have higher emissions per passenger mile—0.56 pound—than a single-occupancy, 50-miles-per-gallon hybrid car (0.4 pound). Intercity highway buses do much better.

The irony is that public transportation poses challenges for a low-footprint personal carbon budget. Most public transportation relies on fossil fuels in the form of diesel buses and trains and conventional electricity grid generation. Public transportation offers huge theoretical efficiencies for energy use and corresponding emissions—if every seat is occupied. But no public transportation system can operate a schedule with every seat occupied all the time. Buses and trains could match the 0.12 pound of CO_2 per passenger mile of the subway system if their seats were fully occupied. But with average occupancy rates of 25 to 50 percent, these forms of transit can't beat the carbon emissions of an electric vehicle.

DOES WORK-RELATED TRAVEL COUNT?

Counting individual carbon emissions poses accounting problems, and accounting for work-related carbon emissions may be one of the toughest accounting issues. In some cases, your work-related carbon emissions clearly belong on some other individual's carbon tab, so you shouldn't count them. For example, if you are the operator of a coal-fired power plant, the carbon emissions of the plant probably belong to the individual consumers who are buying your power, not the operators. But in the more common situation of business travel, no one else is going to count the carbon emissions of your plane trip. Whether you should count those emissions on your own footprint may depend on whether you have a choice about making the trip. Much business travel is not essential; the word *boondoggle* was invented to describe pleasure travel masquerading as a business expense. If you want to be a leader on climate issues, you can model climate-responsible business behavior by using videoconferencing for meetings and virtual conferences. But if you really will be fired from your job if you refuse to make a particular business trip, you can probably count those emissions as the emissions of your business enterprise, not your own—just like the coal power plant operator. I treat all business travel as part of my own footprint, but I am unusually fortunate: as a tenured university professor, I won't lose my job or suffer a pay cut because I refuse to fly to a meeting or conference.

I am a big fan of public transit as a low-impact, stress-relieving way to get to work. You can't read, chat on the phone, or catch up on work behind the wheel of a car (or a bike, for that matter). The Tappan Zee Express bus was my daily commute option for years and remains my backup for days that I can't bike or motorcycle to work. Public transit benefits everyone by reducing traffic congestion. But once I started counting carbon in my daily life, I found myself hopping on the bus or train less often. Commuting on my e-moto or electric vehicle counts as zero carbon, while the 30-mile round trip on the highway bus saps a few bucks from my carbon budget—4 pounds, to be exact.

FOOD MILES AND FOOD SMILES: A MEDITERRANEAN CARBON DIET APPROACH

Keeping track of your food-related carbon impacts may be the toughest part of keeping a carbon budget. Few people keep track of everything they eat, unless they are on a very strict health diet. Carbon accounting issues abound. Whose carbon tab does the roast beef served at the business meeting go on? How about when you are invited to dinner and your host serves lamb chops? Even if you could keep track of everything you eat, estimates of the carbon impact of different foods vary widely, making an exact accounting difficult. And unlike the carbon impacts of burning gasoline in your car or gas or oil in your furnace, where we can attach a number to the direct emissions of our activities, the carbon impacts of food do not occur when we consume it, but from the manner in which it is produced. So any carbon accounting for our food consumption is based on a different system of measurement than simply calculating the greenhouse gas emissions that result directly from our activities; it involves estimating the life cycle emissions of various food production practices around the world. Even for a given sort of food, greenhouse gas emissions can vary wildly according to the production method and the methodology used to calculate the impacts. The carbon impacts of shrimp production, for example, vary from 11 pounds to 2,205 pounds of CO_2e per pound. Wild-caught shrimp have the lowest impact. The highest emissions impacts for shrimp are for Far East and Indonesian shrimp farms because of the greenhouse gas impacts of clearing mangrove forests to make room for shrimp farms.

FAMILY MATTERS: HOW DO CHILDREN COUNT FOR YOUR CARBON FOOTPRINT?

No discussion of household emissions is complete without considering who makes up your family household. For the vast majority of people, the concept of home and family includes children. So how does having children affect your household carbon footprint? British researcher

Kimberly Nicholas caused a stir in 2017 when she published a chart showing high-carbon-impact personal choices and indicated that the impact of having one child in the developed world was 58 tons per year—far more than air travel or meat consumption. This 58-ton figure was misleading however, as it did not correlate to greenhouse gas emissions in any actual year. Rather, the researchers calculated the emissions not just of the child, but of all her descendants for thousands of years, and then attributed a portion of these emissions to current time—not a meaningful comparison to the decision to take a flight and cause fossil fuel emissions immediately.

There is no doubt that the climate crisis is related to population: the sheer number of people on this planet burning fossil fuels and consuming food and energy is what throws the climate system out of balance. However, any attempt at direct population control is ethically fraught and susceptible to abuse. Reproductive freedom is a basic human right, and it includes the freedom to choose to have children as well as the freedom to choose not to. Studies show that the best way to limit population growth in the developing world is to educate women and provide access to birth control so that the choice not to procreate is available.

As far as climate-responsible family planning in the developed world goes, aiming for zero population growth is a defensible option. Two children per couple is slightly below zero population growth, so having two children should not lead to climate guilt. Having children gives you a stake in the future climate and more incentive to act to control emissions. And as a carbon accounting question, if you keep track of carbon emissions per person, having children in your household actually reduces the per capita carbon impacts, as the emissions associated with heat and light are shared among more people. My approach to carbon accounting assumes that every human being is entitled to his or her footprint.

One thing about food impacts is clear: meat and dairy protein generally has much larger greenhouse impacts than vegetable-based protein. It is also clear that, because of the methane from intestinal gases, meat from ruminants such as cattle and sheep has much larger impacts than meat from nonruminants such as pigs and fowl. Meat production also requires much greater land disruption per gram of protein than vegetable protein, as far more grain goes into feeding the animals than it would take to feed people directly. The need for open land to grow grain for livestock adds pressure to clear forests. Deforestation adds to global warming, as the carbon stored in cleared forests is released into the atmosphere as carbon dioxide and methane. So the carbon impacts of that hamburger on your plate may vary considerably depending on how and where the cattle were raised, processed, and transported.

Many people will tell you that the only responsible diet is one that foreswears all animal products—become a vegan. There are certainly reasons to do so: most livestock in the United States is raised under conditions of unspeakable cruelty and environmental atrocity. Hog production facilities consist of tens of thousands of hogs confined in pens so small they cannot turn around during their lifetime. The millions of gallons of manure from these facilities poison ground and surface water; the manure receives less treatment than human sewage and is spread on fields at rates that ensure runoff into groundwater and streams. Cattle and poultry operations often are not much better. These animal production facilities do not merit being called farms: they are meat factories.

But my Mediterranean diet approach to climate and diet is nondogmatic. Humans have been omnivores throughout their existence; any approach to climate mitigation that relies on human beings abandoning their essential nature seems unlikely to succeed. The question need not be whether to "eat no meat or dairy" if it can be posed as "How much meat and dairy are consistent with a sustainable carbon budget?" The Mediterranean diet approach works well here: a diet that is primarily plant based but that includes small amounts of meat as a garnish and occasional meat-based meals as a special treat for celebrations or festival days. And not all meat eating need be inhumane; if you eat meat only as a rare carbon luxury, you can afford to spend more to buy only free-range, pasture-raised beef, chicken, or pork.

So how much meat eating is possible on a 4-ton direct carbon budget? Most online calculators use conversion factors of 28 pounds of CO_2e per pound of beef or lamb, about 12 pounds per pound of cheese, about

4 pounds per pound of chicken or pork, and just 0.32 to 6.36 pounds per pound of fish (1–4 pounds for tuna), with a wide variation depending on the species and whether it is farmed or wild-caught (farmed fish have much higher impacts). Using these numbers, if you set aside one-eighth of your carbon budget (1/2 ton) to cover meat eating, you could still have about 12 pounds of beef or lamb per year (1 pound a month), 25 pounds of chicken or pork (2 pounds per month), and 25 pounds of dairy products—hardly an austere diet. But a 2018 study published in *Science* magazine estimates that the median impacts of meat production, accounting for land-clearing activities, waste, processing, and transport, are five to ten times larger than the published numbers.

My own Mediterranean diet approach is a compromise. I consciously avoid high-impact foods like beef and lamb, eating well less than a pound a month of these red meats. It's not practical to keep track of absolutely everything you eat in your carbon budget, so I don't try to add up the ounces of chicken or pork sausage we add to Mediterranean dishes, or the wild-caught salmon. In a typical week, we might eat purely vegan dinners three nights out of seven, enjoy some wild-caught salmon filets one night, have a poultry-based meal one night, and have cheese- or sausage-garnished Mediterranean dishes the other two nights. Instead of beef in chili, I use vegan meat substitutes or ground turkey. Instead of chicken in the curry, I use sweet potatoes and chickpeas. A homemade minestrone soup is perfect on a winter weekend afternoon—and is nearly entirely vegetable based. But a little bit of pancetta and beef stock make it richer and more filling. Another great (and simple) winter dish is Portuguese "green soup," or *caldo verde*, which is just kale, onions, and potatoes and a little olive oil—but it is much more fulfilling when you start with chicken stock and add a garnish of sautéed sausage. Garbanzos and chorizos is a classic Spanish comfort food full of vegetables and contains a small portion of pork chorizo sausage per serving.

Cuts of red meats like beef and lamb are "festival day" treats. To be conservative, I count the carbon impacts of these high-impact foods against my carbon budget whenever I choose to eat them, whether I buy the hamburger myself (rarely) or simply choose to have one of the burgers at the office picnic. Buying local produce and free-range meat at local farmers' markets in season helps minimize the transportation-related carbon impacts of your food consumption and helps to support local agriculture. Meeting regional farmers and feeling that connection to the land that produces your food add to the pleasure of buying, cooking, and enjoying food.

But remember that reducing beef and lamb consumption will have a much bigger impact on your annual carbon footprint than carbon emissions related to food transportation. Keeping track of these high-impact foods will reduce your consumption, since it will force you to think about whether that particular steak, chop, or hamburger is really worth the carbon impacts. Wild-caught fish and farmed mollusks like mussels and oysters have much lower greenhouse impacts per serving than beef or lamb. Farmed salmon also has relatively low greenhouse gas impacts, though this form of aquaculture has other environmental impacts that lead most environmental groups to recommend avoiding it. Freshwater farmed fish and farmed shrimp have greenhouse gas impacts comparable to beef. Chicken and pork fall somewhere in between.

The following chart ranks common meat-based food protein sources per 40-gram serving (about 1/2 pound for most meats) and calculates how many servings per year you could eat if you devoted one-eighth of a 4-ton carbon budget to food protein. Since 40 grams is close to the 50-gram FDA recommended daily protein intake, you would need about 365 servings per year of any one of these foods to meet nutritional requirements. According to this chart, you could eat unlimited amounts of mollusks and wild-caught fish (over a thousand servings) without blowing your carbon budget and could comfortably stay within this budget eating chicken or pork (hundreds of servings), but you could not meet this budget eating primarily beef, farmed freshwater fish, or farmed shrimp. Few people would eat only one sort of protein, so a responsible diet can include beef and lamb as a treat, if it relies primarily on legumes, saltwater fish, or poultry as its primary source of protein.

Not all land is suitable for crop production, with its high water, fertilizer, and soil needs. Not all animal husbandry is cruel and environmentally destructive. When we visited the Azores Islands in the middle

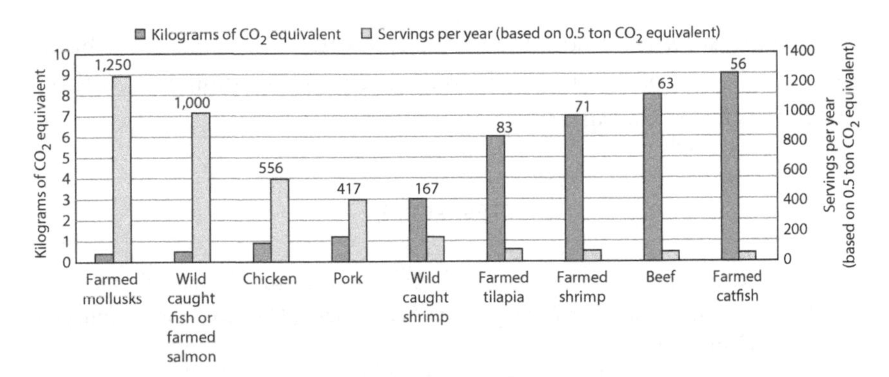

Carbon footprints of meat-based protein sources (not including transportation or processing impacts).

Data from Ray Hilborn, Jeannette Banobi, Stephen J. Hall, Teresa Pucylowski, and Timothy E. Walsworth, "The Environmental Cost of Animal Source Foods," *Frontiers in Ecology and the Environment* 16, no. 6 (August 2018): 329–335.

of the Atlantic Ocean in 2014, we saw an island economy based on cows pastured on the grasses growing in the rocky, hilly soils that were not suitable for grain or legume crops. The cows seemed to be humanely treated, left to pasture unperturbed. It seemed like the right place to have a locally butchered steak in a local restaurant. While veganism will reduce your food footprint most dramatically, there is still room in a responsible carbon budget for occasional meat-based protein. And all the farmed clams you care to eat.

ELECTRICITY, HEAT, COMMUTING, AND FOOD WRAP UP

With a little planning and no major dislocations, you can get your carbon footprint for electricity, heat, commuting, and food down to about 3 tons in a 4-ton annual direct carbon budget. Renewable energy contracts and rooftop solar can zero out your electricity footprint. Electric heat is a relatively inexpensive retrofit, though it adds to your heating bill. EVs are an attractive choice for getting to work for those who can't start bicycling. By making careful food choices, you can keep your food footprint under 1 ton without completely giving up poultry and meat. You can take each of these measures right now. If you can keep your emissions for the basics of life down to 2 or 3 tons, that leaves some room for guilt-free, carbon-emitting fun activities as well.

REFERENCES

Adams, Nate. *The Home Comfort Book: The Ultimate Guide to Creating a Comfortable, Healthy, Efficient, and Long-Lasting Home.* Scotts Valley, CA: CreateSpace Independent Publishing, 2017.

Casey, Allison. "Program Your Thermostat for Fall and Winter Savings." *Energy.gov*, October 9, 2012. https://www.energy.gov/energysaver/articles/program-your-thermostat-fall-and-winter-savings.

"Costs and Benefits of Air Sourced Heat Pumps." *Energysage*, May 2, 2019. https://www.energysage.com/green-heating-and-cooling/air-source-heat-pumps/costs-and-benefits-air-source-heat-pumps/.

"Heat Pump Systems." *Energy.gov*. https://www.energy.gov/energysaver/heat-and-cool/heat-pump-systems.

Jenkins, Jerry. *Climate Change in the Adirondacks: The Path to Sustainability*. Ithaca, NY: Cornell University Press, 2010.

Lloyd, Donal Blaise. *The Smart Guide to Geothermal*. Masonville, CO: PixyJack Press, 2011.

MacKay, David J. C. *Sustainable Energy—Without the Hot Air*. Cambridge: UIT Cambridge, 2009.

New York State Energy Research and Development Authority. "Heat Pumps." https://www.nyserda.ny.gov/Residents-and-Homeowners/Heat-and-Cool-Your-Home/Heat-Pumps.

Thomas, Dirk. *The Woodburner's Companion*. 3rd ed. Chambersburg, PA: Alan C. Hood, 2006.

U.S. Environmental Protection Agency. "Greenhouse Gas Equivalencies Calculator." December 2018. https://www.epa.gov/energy/greenhouse-gas-equivalencies-calculator.

U.S. Environmental Protection Agency. "Methodology for Estimated Energy Savings from Cost-Effective Air Sealing and Insulating." 2011.

CARBON DIARY

April 2016

Far from being the cruelest month, April in New York is a celebration of spring and rebirth, and, for the environmentally minded, a celebration of Earth Day. Or, in my case, Earth Week.

For several years now, my Earth Week celebration usually includes going without fossil fuels for the entire week: because I know that we all have to learn to live without them in the not-too-distant future, and because I know that burning fossil fuels is the thing we do as individuals that harms the ecosystem most. If Gaia is my secular religion, then Earth Week is my week of abstinence, my Passover, my Lent—just to remind myself that fossil fuels are not a gift from Mother Earth, but a theft from her equilibrium.

Going fossil free for a week does not mean giving up all that much, since my low-carbon project has eliminated fossil fuels from most of my daily life. But I still have a natural gas–powered water heater and cookstove. On Saturday, April 16, after a quick Google search, I declared the start to Earth Week and had a discussion with Robin about shutting off the natural gas and taking a hot shower right away with the water that was already in the water tank before it cooled down too much. Robin is wonderfully supportive. Actually, I think she managed to be traveling during all of my previous fossil-free Earth Weeks.

So I got up, shut off the gas to the water heater, took a hot shower, made a waffle breakfast that couldn't be beat, and set off to enjoy a fine and fossil-free April day. We went hiking Saturday and Sunday, taking the electric car to the trailhead. I got to work by kayaking and biking on Monday and Thursday. I rode my electric motorcycle to work on

Paella over an open fire.

Photo courtesy of Karl Coplan.

Tuesday and Wednesday, and I charged it with the solar panels on the roof. I took the electric car to work on Friday because of the forecast for rain. Instead of running the dryer, we hung all our clothes out to dry in the sun. We found creative ways to cook all week without the gas stove: we spent Saturday night on our sailboat at her winter berth and cooked supper on an alcohol stove Saturday. Sunday evening was perfect for a barbecue on a charcoal grill. We pulled out the electric induction hot plate on Monday and Tuesday (very efficient), and a crock pot powered with solar panels Wednesday. On Friday, I dug up my twig-fueled camping stove to cook a pot of chili for dinner. Our experiment with a solar oven didn't work so well on Saturday, April 23, so we threw the chicken sausage onto the twig stove, too. Then on Sunday we threw a paella party to celebrate the end of Earth Week, and we cooked it the traditional Spanish way, on a paella pan rack over an outdoor wood fire.

Ah yes, the lack of a water heater for the week. No, we didn't give up personal hygiene; we made do with sponge baths in warm water heated with the electric induction heater. One benefit of doing entirely without fossil fuels for a week is that it's a reminder of how one's life still absolutely depends on fossil fuels—in my case, hot water, the gas dryer, trips more than 30 miles from home, and that fine-tuned temperature control for cooking that only a gas stove seems to have. At least substitutes are in sight for the hot water, the dryer, and the longer-distance travel: I could replace my gas water heater with an electric one at any time (but the extended warranty on the one I have lasts to 2030!), ditto for an electric dryer, and an affordable electric Tesla with the range to go to the mountains or the seaside should be within reach in a year or two.

So we kept our weekend travel close to home this month, enjoying hikes in Harriman State Park and High Tor State Park—both within range of the electric Smart car. One Saturday, we went for a cruise through Bear Mountain State Park on our electric motorcycles. At the end of the month we did drive up to our cabin in the Adirondacks in our Prius to do some early spring hiking and to set up Robin's upstate beehives.

The month was just mild enough that the gas heat never came on in our house, though I did spend a weekend morning gathering some more downed branches from the backyard to supplement our depleted woodpile for the woodstove. My total carbon footprint for the month was just 158 pounds—my lowest monthly footprint yet since I started keeping track last September. The biggest item for the month was natural gas for the water heater, dryer, and stove, totaling around 80 pounds. At this rate, I might keep to an annual carbon footprint of less than 1 ton—sustainable under almost any definition—but that would not leave any carbon budget for travel this summer or trips to the mountains.

9

HAVING FUN ON A CARBON BUDGET

The last chapter examined strategies to reduce or even eliminate the direct carbon footprint of daily life—running your home, getting to work, sustenance. But a life fully lived involves more than these basics. No matter how rewarding we find our careers, the things we choose to do for their own sake are part of how we express our individual selves and restore our energy. Whether you call it leisure, recreation, or play, this unobligated time plays an important role in supporting well-being in those societies (such as the contemporary United States) lucky enough to allow for significant leisure time. Whether it is a regular jam session, soccer league, or cultural enjoyment, these activities help us define ourselves, build family and community bonds, and make us whole.

The good news is that many of these activities are by their nature low-carbon pursuits, such as nonmotorized sports and hobbies. The bad news is that many other leisure activities—particularly vacation travel—have become dependent on fossil-fueled mobility. But if we manage to save 1 ton of a 4-ton carbon budget for these pursuits, there is still room for some fossil-fueled fun in our lives as the overall energy economy transitions, as it must, to zero net emissions.

Since your choice of leisure activities is the essence of individual autonomy, no one can tell you the "right" way to spend your leisure time. What is "right" is whatever you find personally rewarding and expresses your individual self. So what follows is just a set of suggestions, ranging from no-carbon pursuits to low- or moderate-emissions pastimes that can still fit in a 4-ton direct carbon budget. Consistent with a nondogmatic "Mediterranean diet" approach to a low-carbon lifestyle, I will try not to let my own values dictate what is the right way to spend leisure time; as long as it fits, spend your carbon allowance however you want. Living with a carbon budget is itself a value expression though, and is probably inconsistent with a high-spending conspicuous-consumption lifestyle.

THE BEST THINGS IN LIFE ARE CARBON FREE

Too often we treat our leisure activities as a form of conspicuous consumption, spending money and consuming resources beyond the necessities of life to make a statement about who we are. The late-eighteenth-century sociologist Thorstein Veblen coined the terms *leisure class* and *conspicuous consumption*, both of which he criticized as forms of social decadence and class stratification. In the United States today, leisure is no longer the exclusive province of coupon-clipping rentiers. But many people still spend money on vacations and leisure activities so they can brag and bask in the envy of their peers. However, consistent with an approach of conspicuous nonconsumption—making a statement about our values by what we choose *not* to consume—we can start with a list of personally enriching activities that require little or no carbon emissions.

Having a hobby or artistic pursuit outside of work enhances well-being and happiness, and this sort of personal development need not involve travel or carbon emissions. You can learn carpentry, or to play a musical instrument, or to paint, or to grow vegetables. Slow food instead of fast food builds social and family bonds; you can learn to cook great food not because you have to but because it is a way of sharing your love—and a meal—with friends and family. Creative pursuits like these give the joy of adding something beautiful to the world. Learning new skills also exercises your mind and, like physical exercise, improves cognition. Many communities sponsor adult education programs with courses in music and other creative arts.

Some of the most enriching life activities are those that connect you to family and community. You can join a book club, take up a new sport, or rediscover an old one and join a sports league. Not feeling particularly athletic? Not all sports require great athletic conditioning; club sailing and bowling come to mind as social sports that connect you to your community and give you the kick of the occasional strike or good finish. Running clubs and bicycling clubs sponsor fun runs and rides that don't require marathon fitness.

What about motorized recreation? Can RVs of any sort be consistent with a sustainable carbon footprint? Motorized "recreation" seems like a contradiction in terms because re-creating your body usually means physical exercise just as re-creating your soul requires a break from the discipline of work, whether or not you work at a desk job. Many people are restored by outdoor thrills that don't involve physical labor. Even though I favor physical exercise, I have on occasion set off for an electric motorcycle ride on the wooded, winding roads of Harriman State Park. As with everything, from heating your house to commuting to work, how this fits into your carbon budget depends on how you do it and how much you do it. Moderation is the key. Most gas-powered motorcycles will take you farther on a gallon of gas than the typical car. Jet skis, four-wheelers, and snowmobiles, on the other hand, don't generally do as well; a typical jet ski or snowmobile will burn through 4 gallons of gas in 1 hour (though off-road vehicles do a bit better). That's comparable to driving an SUV for the same amount of time, and at that rate your entire 1-ton carbon fun budget might only be good for 25 hours of motorized recreation, even if you do nothing else.

THE CARBON FOOTPRINT OF STUFF

I do not try to keep track of the carbon emissions embedded in the consumer goods I buy. There are too many variables involved; different kinds of stuff have vastly different footprints, and accounting for durable goods poses issues for allocating the carbon embedded over their useful life. When I started tracking my carbon footprint on a daily basis, online calculators made no attempt to keep track of the carbon emissions associated with purchases. For my budget purposes, I have treated these emissions as part of my indirect footprint.

But the carbon emissions involved in the manufacture of consumer goods can be substantial, especially for larger, heavier goods such as automobiles. Indeed, the carbon emissions involved in manufacturing a hybrid car can be nearly as large as the emissions from the fuel used to run it during its lifetime. Carbon calculators such as carbonfootprint.com now calculate a footprint based on various categories of consumer spending. One thousand dollars' worth of clothing or televisions equates to about a third of a ton of CO_2e, while the same thousand dollars' worth of furniture, computer equipment, or books is equal to closer to one-half ton of CO_2e. Avoiding consumerism is the best way to limit this impact; you can use goods and clothing until they wear out and consider buying used goods rather than new.

You can always share your passions with your spouse and children, since the investment you make in your family costs nothing and will pay off in generations of good memories. Participating in community cleanups and neighborhood events will also pay off in an increased feeling of social connection. You probably don't need to read a book about reducing your carbon footprint to convince you to engage in these rewarding downtime activities; chances are you already enjoy at least one of them. The bad news is that Americans' favorite leisure activity is being a "couch potato": nearly half of our leisure hours are spent in front of the television. At least the direct greenhouse impact of this activity is relatively modest: even with the biggest, baddest plasma television on the market, you will only use about 500 kilowatt-hours of electricity in a year of average TV watching—equivalent to about 500 pounds of CO_2 even if you are not buying carbon-free electricity. That's less than burning 25 gallons of gasoline over the year to spend leisure time outside the home. But watching all those advertisements on television will probably make you buy more stuff, which has its own carbon footprint.

TRAVEL AND TOURISM ON A CARBON BUDGET

When we are not watching television, we like to travel. Travel broadens our horizons and opens our minds, and (apart from conspicuous bragging rights) it gives us new stories to share with coworkers and friends on social occasions. But it takes energy to move our mass around over long distances, so leisure travel can easily swamp our carbon budget when it involves long distances. Giving up routine leisure air travel is often the hardest part of committing to a carbon budget.

The United States Travel Association estimates that three out of four travel trips in 2017 were for leisure. Many of these leisure trips are by air; three out of four Americans in a survey stated that they went on a vacation flight at least once a year, and 11 percent reported taking a leisure flight once a month. (Still, on a typical Memorial Day weekend, ten times more Americans will drive to their weekend getaway than will fly.) All that fossil-fueled travel takes its toll on the planet's climate system, as a typical round-trip plane flight will add between 1 and 2 tons of CO_2e to the atmosphere. This makes the typical two-leisure-flights-per-year travel pattern unaffordable on a carbon budget. Let's look at ways to enjoy the benefits of travel while sticking to a carbon budget.

Not all horizons need be distant. On the ocean, the horizon from eye level is only about 5 miles away. There is plenty to discover over the near horizon, with little carbon cost. Longer distances are still possible, too, on a carbon budget, if you are in a position to trade time and convenience for these more distant horizons.

Locatourism: Discovering Your Regional Resources

It is useful to think of the reasons that make travel and tourism pleasurable (apart from the bragging rights of going where your friends haven't been). Perhaps it is the recreational opportunities: boating or swimming, walking at the beach, or finding solace and solitude in the mountains. Or perhaps it is the excitement of new cuisine or music or theater performances, or casino gambling. How far do you need to go to satisfy these yearnings?

We naturally assume that "new and different" means distant; the word *exotic* means "from another country." But as long as your travel is powered by fossil fuels, its carbon impact is going to be directly proportional to the distance you go. This is true whether you travel by plane, train, bus, or automobile. So the most effective way to reduce your personal travel footprint is to stay closer to home. Be a tourist in your own community: discover the art, culture, history, recreational resources, and cuisine of the place where you live. You can start with your state's own tourism and recreation websites. Chances are there are museums, historic sites, beaches, and trails you have skipped over while setting your sights on more distant horizons. Just as you have learned to appreciate the sense of community and connection that comes with buying food from regional farmers at your farmers' market, you can gain a sense of connection to place by discovering your own region's history. Instead of flying to Napa Valley (at least a 1-ton round trip), look for a local winery and take a tour. The French use the word *terroir* to describe the taste for wine from one's own region; the word refers to the earth itself in which the grapes are grown. Thanks to the wonderful ethnic diversity of our country, you can sample exotic cuisine in almost any city, so you can try going to a restaurant in an immigrant neighborhood you might normally avoid and can travel internationally without crossing any borders. You can combine your locavorism with locatourism: develop a taste and appreciation for local attractions. If your inclination is more toward Vegas than Paris, legalized casino gambling is now sufficiently common that you can find a casino in day-trip range.

Once I had my little electric car, which would carry me, Robin, and the dog around for 60 to 80 miles round trip on one charge, we started planning our weekend trips within that range. These trips were "free" on my carbon budget as long as I charged my car with renewable electricity. Living in the fringe suburbs as we do, that electric range can take us everywhere from the Shakespeare festival at Boscobel in the Hudson Valley, to the vistas of the not-so-high peaks of the Ramapo Range in Harriman State Park, to all the excitement and culture that Manhattan has to offer. There are even two lift-served ski areas in our EV range (but only if we shiver without heat there and back). Our EV is not quite enough to get us to an ocean beach in the summer, but newer EVs have plenty of range for that, too. And our once-a-month or so trip in the hybrid to our Adirondack Mountain retreat fits comfortably in my carbon budget. One thing I discovered while exploring more regional outdoor activities is that you can often find more solitude in a regional state forest or park than in a destination park. As a Colorado friend once quipped to me about Rocky Mountain National Park (an air travel destination for me), "the problem with Rocky Mountain is that you have to share your solitude with lots of other people."

LOW-CARBON TRAVEL BEYOND THE WEEKEND GETAWAY

Face it: if you have wanderlust, sooner or later you will want to travel beyond EV range, if only to hike higher peaks, escape from the oppressive heat or cold of your home climate, or experience other cultures. But of course, travel beyond day trips or EV range gets to be more carbon intensive, and discretionary air travel is where otherwise climate-conscious Americans rack up outsized carbon footprints. I can't tell you how many committed environmentalist acquaintances of mine start a conversation by recounting all the places they have traveled by air in the past year, while remaining convinced that their carbon footprint is below average because they shop at a local farmers' market and drive a hybrid. So the key to low-carbon travel is to avoid air travel at (almost) all cost. A low-carbon family vacation will most likely take the form of a very retro road trip in the family car or some other form of slow travel.

Take a Road Trip

It seems odd as an environmentalist to be extolling the virtues of the old-fashioned, fossil-fueled freedom of the open road in your personal automobile. But if you have a week or two, a family road trip is probably the least harmful, realistic way to get beyond your regional backyard. Pack a family of four into a car (or even a small truck or SUV), and mile for mile, you will be responsible for fewer direct carbon emissions than if you boarded that same family onto a jet plane. Your relative carbon emissions will be even lower because you would not dream of driving nearly as far as you would be willing to fly.

Here is what to love about the automotive vacation (at least when compared to air travel). Your car almost certainly has more legroom than the average coach-class seat on an airplane. Rather than being treated as a passive form of cargo to be shipped from one airport to another, you can control your travel, deciding when to leave and when to stop. You have an unlimited choice of roadside food, including local diners and rest stops that make you feel that you really have traveled beyond the food choices of your local shopping mall. And bad weather in Chicago will very rarely end up delaying you by eight hours. You do not need to bare your person and possessions to the TSA to be allowed to go anywhere. You will probably spend a good deal less for your travel day, and there will be no extra charge to bring your stuff with you. And you can carry on most liquids in unlimited quantities.

One reason the road trip is becoming more popular as a family getaway is that, between the enhanced security required after the 9/11 attacks and the market pressures on airlines to cut basic ticket prices, air travel has become simply unpleasant. Those of us who have largely given it up hardly miss it. Get in your car, and you can make your travel plans at the last minute. Food and lodging—even your destination can be spontaneous. Choose a hotel or campground for the night when you feel like stopping, and see a place that looks interesting. Interact with people who are actually from the place you are traveling to, rather than tourist-oriented hotel and airport personnel. Because your travel can be unplanned, you (and your family) will have more memorable stories to tell, because it is the unplanned things that make travel more interesting.

Between traveling to and from the airport, arriving two hours early for security, waiting for ground transportation at your destination, and recovering from hours crammed into a small personal space with low oxygen, air travel kills an entire day wherever you may be bound. For the majority of people living in the United States, a long day's road trip by car can take you to your choice of stunning mountain landscapes, deserted ocean (or Great Lakes) beaches, world-class cities, or an international border with a language and culture other than English on the other side.

Slow Travel

There is a direct relationship between transport speed and carbon emissions. The Concorde supersonic passenger plane, when it was operating, generated five times more greenhouse gases per passenger mile than subsonic aircraft. Air travel generates about four times more greenhouse gas emissions per passenger mile than four people sharing the average automobile, while even four people in the average automobile generate twice as much greenhouse gas emissions as the same four people on an intercity bus.

Long-distance travelers on a carbon budget must confront this tradeoff between time and harm to the planet. The cost of low-carbon travel can be measured in time. Just as there is a movement toward slow food, there is a movement toward slow travel.

What does slow travel mean? Compared to flying, even driving to your destination is a form of slow travel, especially if you stop to learn about the part of our nation air travelers derisively refer to as "flyover country." The slower you go, the more of the world you will experience along the way. Slow travel gives truth to the old saying "it's not the destination, it's the journey." The slower you go, the more connected you will be to the land and the people on the way. Even when you travel by car, the interstate and your windshield isolate you from the surrounding land and communities nearly as much as the 30,000-foot elevation of an airplane. If you really want to travel to a place rather than teleport there, try traveling slowly and without a ceiling or walls around you.

SLOW TRAVEL BY SAILBOAT?

In July of 2013, Robin and I were climbing Gros Morne Mountain—the highest peak in Newfoundland—looking down at the fjords and the still-melting snowbanks. It had taken about a week to sail to Newfoundland. We talked about low-carbon travel, and Robin said pointedly: "You can't just tell everybody they need to sail to their vacation." She's right. While sailing on the ocean now feels safer to me than riding my motorcycle or my bicycle, or even driving a car in traffic, that comfort comes from a lifelong passion for messing about in small sailboats. And it's not just a rich person's pastime; once you get beyond the U.S. coast, you see many families with children on sailing sabbaticals. We have sailed our boat to Barcelona, Lisbon, Senegal, and Guadeloupe, from icebergs to palm-studded islands.

But if you are not quite ready to substitute the mercy of the seas for the mercy of the airlines, you can still get to a place like Newfoundland, at least, without leaving the ground. Driving from New York to the Newfoundland Ferry in Sydney, Nova Scotia, could easily be done in the same week it took us to sail there, and, unlike us, you could sleep every night on the way.

If palm-studded beaches are more to your liking, a road trip plus a short flight can take you to the Bahamas on the east coast or to Baja on the west. And if you do get the sailing bug, there are plenty of adult sailing schools and bareboat certification programs that will give you the basic skills you need.

Slow travel requires time, and for some forms of travel, a modicum of physical fitness and adventurousness. For those who can arrange the time and the physical conditioning, taking one of the great hikes, such as the Appalachian Trail or the Pacific Crest Trail, will give you stories and bragging rights that no patron of a prepackaged air travel ecotour can claim. As would riding a bicycle across the continent. For the aquatically inclined, there are long-distance paddling routes in many parts of the country. If the prospect of that much exercise puts you off, a cross-country motorcycle trip could also be an adventure of a lifetime. Several enthusiasts have made the transcontinental trip on electric motorcycles.

WHY NOT AMTRAK?

I am absolutely enthralled with train travel—I have taken Amtrak from New York to Colorado twice, and to Chicago, Pittsburgh, and Charleston as well. The train is much more pleasant than the bus or the plane, and the dining car still seats you with strangers, so you end up sharing stories with people you might never interact with otherwise. But the carbon footprint of U.S. rail travel is just not that good; a study by the Union of Concerned Scientists puts Amtrak's emissions at about 0.37 pound of CO_2e per passenger mile. This is not much better than flying the same distance. And that figure is for a coach seat; if you book a roomette for an overnight train trip, your emissions will be higher than if you flew. If you are looking for low-carbon slow travel options, a road trip with at least two people, or the Greyhound bus, beats the romance of rail travel hands down, at least until long-distance rail travel in the United States is electrified.

Unless you are planning a trip down the Pan-American Highway, intercontinental slow travel means going to sea. There were international travelers long before there were jet aircraft. But don't think of riding a cruise ship as a low-carbon option. Even though cargo ships in general are much more energy efficient than most other transport, modern cruise ships have carbon footprints per passenger mile that are as bad as air travel for the same distance. This is because a cruise ship is carrying many more crew members per paying passenger and because the ships are driven faster than their most efficient speed. Greener cruise ships may be on the horizon, but we are not there yet.

Until carbon-friendly cruise ship bookings are available, it is still possible to book a room as a passenger on a container ship, and that is likely to be an unforgettable experience as well. As of this writing, Maris Freighter Cruises is offering transatlantic and transpacific bookings. Or, if you are really adventurous, look for a spot as unpaid crew on a sailboat; private skippers are often looking for extra hands, and some are willing to take inexperienced crew. There is a website—crewfinder.com—that links blue-water sailors with volunteer crew members. If you sail to your destination, your travel carbon emissions may be close to zero.

AIR TRAVEL AS A TREAT

As we have seen, a 1-ton carbon budget for fun does not allow for much air travel, if any. The carbonfootprint.com calculator shows that you would use that 1 ton with one round-trip flight to and from New York and Los Angeles, or to Paris, or to Cancun. Many other estimates put the carbon impacts of flying much higher—perhaps twice as much. But consistent with a nondogmatic, Mediterranean diet approach to your carbon footprint, one need not say "never" to air travel. Like meat-based foods, it will have to be a rare luxury rather than a routine escape. No one should feel guilty about taking a once-in-a-lifetime trip that truly expands his or her horizons. There are also some events for which your family won't forgive your absence based on your adherence to a carbon budget.

Of course, if you find yourself taking that once-in-a-lifetime trip on an annual basis, then it is not really once in a lifetime. My own rule of thumb is that I am not willing to fly someplace unless I am willing to spend two weeks at the destination. But I will still fly to the wedding or funeral of a close family member. It may mean I blow my carbon budget in a year, and I will try to make it up in the next year to average it all out.

REFERENCE

Union of Concerned Scientists. *Getting There Greener: The Guide to Your Lower-Carbon Vacation*. December 2008. https://www.ucsusa.org/sites/default/files/legacy/assets/documents/clean_vehicles/greentravel_report.pdf.

CARBON DIARY

May 2016

May is when spring really arrives in the Northeast. The trees leaf out and the cold rains of April fade into the pleasant warmth of the May sun. No need for either heat or air conditioning this month (even if we had an air conditioner), and I can make my zero-carbon paddle-and-pedal commute to work several times a week without fear of cold winds. May is also the start of sailing season on the Hudson, so we commissioned our sailboat and brought it back to Nyack from its winter berth in Haverstraw. By the end of the month, we could spend the night anchored off of Hook Mountain State Park in our own floating low-carbon waterfront home.

But May turned out to be a month of travel, too, which added to our carbon tab for the month. We awoke on May 1 at our cabin in the mountains and hiked up Moxham Mountain to see the last patches of snow melting on Gore, then drove back downstate for the workweek. Our daughter graduated from Hampshire College this month, so we made the drive to Amherst, Massachusetts. Robin had her high school reunion in Troy, New York, so we drove up there. While Robin reunited with her school friends, I took the opportunity to try the tourist train that runs from Saratoga (near Troy) to our cabin in the mountains, riding the rails to the historic North Creek train station and then going for a long hike on wilderness trails to get from the train station in town to our cabin in the woods. The tourist train was fun and picturesque, running through pine forests and crossing river gorges, but it was pretty much a bust as a low-carbon travel method, since the diesel locomotive pulled two cars with only a handful of passengers. I estimated the ride as 30 person-miles per gallon for my

The America's Cup qualifying races in New York Harbor.

Photo courtesy of Karl Coplan.

carbon footprint. (A person-mile-per-gallon is a measure of how many people a gallon of fuel will move a number of miles. So a 50-miles-per-gallon hybrid with one person in it has an efficiency of 50 people-miles-per-gallon, but if that same car has two people in it, the efficiency doubles to 100 people-miles-per-gallon).

The Americas Cup qualifying races were in New York Harbor later this month, so we made a bicycle outing out of it and pedaled all the way to lower Manhattan to watch the races from the Battery, and then took the ferry to Jersey City and biked back up the west side of the Hudson River to get home. For Memorial Day—and Robin's birthday—we invited a bunch of family and friends for a paella cookout over a zero-carbon wood fire in the backyard.

Then on May 30, I got on a train for North Carolina to attend the Waterkeeper Alliance annual meeting and conference in Wilmington. On the 16-hour train and bus ride south I did some research into how much carbon emissions intercity rail travel saves compared to flying the same distance. The results were a little disappointing: the CO_2 emissions of rail travel are less than air travel, but not that much less. Depending on how you account for air travel—and depending on whether your train is diesel or electric—the rail travel carbon emissions may be only 20 percent less than the emissions of traveling by air. Long-distance travel is still a carbon-intensive activity.

With all that travel, the total carbon tab for the month came to about 650 pounds—more than offsetting April's light carbon footprint. But it was still on track for a 4-ton annual carbon budget.

10

MEDIUM-TERM GOAL

Getting to Zero

Keep in mind that no use of fossil fuels is "sustainable" in the strictest sense of the word; even apart from the disruptive impacts of climate change, there simply are not enough easily recoverable fossil fuels for future generations to keeping burning them at the current rate. As mentioned earlier, the Brundtland Commission's gloss on the idea of sustainable development allows for continued use of nonrenewable resources only to the extent that the rate of technological progress would enable future generations to enjoy a comparable standard of living.

In the case of greenhouse gas emissions, any assessment of the sustainability of one's individual footprint depends, then, on an assessment of whether technological progress can find a way to replace fossil fuels, as well as of the time remaining before the global carbon budget to avoid warming in excess of 2°C runs out. Burning any fossil fuels now can only be considered sustainable if technological progress to replace them is proceeding fast enough to beat the clock on catastrophic climate change. Keeping in mind Yogi Berra's famous admonition that "it's tough to make predictions, especially about the future," this chapter will try to assess how long we have to achieve a zero-carbon energy economy and what that might mean for individual mobility and energy use in a carbon-limited future.

HOW LONG DO WE HAVE?

In 2012, Bill McKibben's "Do the Math" *Rolling Stone* article predicted that, without immediate action, the world would blow through the carbon budget needed to stay within 2°C of warming within fifteen years.

Half that time has passed, and we are more than half of the way toward burning through the 562-gigaton carbon budget McKibben proposed (see chapter 1). Keep in mind that although there is near scientific certainty that human emissions are causing significant warming, there is much less certainty about “climate sensitivity”: calibrating exactly how much warming will occur based on a given greenhouse gas concentration in the atmosphere. So carbon budgets for 2°C of warming are expressed in terms of probabilities. McKibben’s budget was based on the emissions reductions needed to have an 80 percent chance of exceeding 2°C of warming. The Intergovernmental Panel on Climate Change (IPCC) uses the more risk-tolerant level of a 66 percent likelihood in determining the remaining carbon budget (see chapter 3).

Based on the 350.org budget, then, the world would only have about six or seven more years to convert to a nonfossil-powered economy. The International Renewable Energy Agency, on the other hand, predicts that, based on the less austere IPCC budget, the remaining carbon emissions budget will be used up by 2038. This forecast takes into account the current rate of renewable energy deployment. The 2015 Paris Agreement for limiting global greenhouse gas emissions adopted a less aggressive goal of zero net emissions by 2050, but IPCC estimates of the reductions needed to limit warming to 2°C point to an earlier deadline for achieving zero net emissions, and they assume that costly “negative emissions” technologies (energy-intensive carbon capture from the air) will be implemented to reach net zero. The 2018 IPCC SR15 report on the measures needed to limit warming to 1.5°C estimates that emissions need to decline by 50 percent by 2030.

No matter which approach you follow, then, the clock is ticking on the fossil fuel economy, both for individuals and nations, and that clock will likely run out no later than 2040—barely twenty years from now. Will zero-carbon emissions technologies be available in time? Will you be ready personally for a zero-emissions energy economy? There are reasons to be cautiously optimistic—with an emphasis on the caution part. We can cover basic human needs with renewable energy, but some of the luxuries we have come to expect in the developed world may not be available in a zero-carbon energy economy.

What a Zero-Carbon Energy Economy Might Look Like

We would all like to believe that there is a "silver bullet" to slay the werewolf of climate change: cheap and plentiful renewable energy that simply outcompetes old-fashioned fossil fuels. You have probably heard some very optimistic assessments of the abundance of renewable energy just waiting to be harnessed. The U.S. Department of Energy's 2012 *SunShot Vision Study* on the potential of solar energy in the United States points out that 0.6 percent of the land area of the United States could supply all of the nation's electricity needs. And a wind power engineer recently asserted, "There is enough wind energy capacity off the east coast to power the entire country." As the nation's preeminent techno-optimist, Elon Musk of Tesla Motors fame, put it in a 2017 speech to the National Governors Association,

> If you wanted to power the entire United States with solar panels, it would take a fairly small corner of Nevada or Texas or Utah; you only need about 100 miles by 100 miles of solar panels to power the entire United States. The batteries you need to store the energy, to make sure you have 24/7 power, is 1 mile by 1 mile. One square-mile.

Problem solved, right? All we have to do is divest from stock in Exxon-Mobil and American Electric Power and the solar and wind grid will build itself. But there are some problems with these back-of-the-envelope energy calculations. The people who live in the Southwest might object to 10,000 square miles of their state being commandeered for power generation. Moreover, all those solar panels would require the mining of huge amounts of rare elements, which would require energy that is currently supplied by burning fossil fuels. Estimates of the "energy returned on energy invested" (EROEI) for photovoltaics vary widely—from 2.4:1 to 38:1. At an EROEI of 2:1, half of the entire lifetime output of a solar panel would be needed to pay back the energy debt of its manufacture. EROEI estimates for wind power also vary widely—from 1.27:1 to 77:1.

Once all this solar and wind generation capacity is manufactured, there is still the problem of the temporal and spatial mismatch between the power generation and the power demand. Put simply, the sun may be shining all day in Arizona, but that does little good for the householder in northern New England who is trying to heat her house with an electric heat pump on a winter night. Moving that power spatially will require an enormous expansion of electric transmission lines, which will require more mining of materials, more construction, and more energy investment. Moving electricity long distances also causes energy losses in the wires, adding to the energy needs. New high-voltage direct current (HVDC) transmission lines offer lower energy losses, but they still require massive construction and materials installation.

Then there is the timing mismatch: solar energy peaks midday, while electricity demand peaks in the evening just as the solar energy drops off rapidly. Wind power is less predictable and does not generally offset the variability of solar power production. Utilities in California, with all its rooftop solar installations, already experience grid management problems because of jagged daily residential electricity demand. Demand drops to near zero in the middle of the day when all that rooftop solar is adding energy to the grid and residential meters are running backwards, and then spikes in the evening when homeowners crank up the air conditioning at the same moment that the sun sets on solar power. This demand curve is called the "duck curve" because it resembles a duck, and it poses severe problems for the grid, because few forms of power generation can be ramped up from zero to peak fast enough to handle the evening spike, and most of those rampable forms of power are fossil fueled.

Elon Musk suggests that battery storage provides the answer to the evening peak and overnight energy problem. But the lithium-ion batteries that Tesla is selling have their own embedded energy bill: the energy storage on energy invested for a lithium-ion battery is about 10:1, meaning that you need to invest 10 percent of the total lifetime energy storage of a battery in making the battery. Tesla recently installed the first utility-grid-scale lithium-ion battery plant in South Australia, rated at 130 megawatts of power capacity and with a storage capacity of 130 megawatt-hours. This is a substantial accomplishment; it allows the utility to substitute battery storage for building new generation capacity to handle the peak hour of demand. But it only covers the peak hour; at 100 megawatts, it will use up the entire battery capacity in about an hour and twenty minutes.

If you were to try to scale the South Australia project up to cover storage for one day of current U.S. electricity demand (10 million megawatt-hours, approximately 100,000 larger), the cost would run into the trillions of dollars, even assuming that sufficient lithium could be mined and the batteries could fit into Musk's 1 square mile.

These calculations are based on current electricity demands. But a zero-carbon energy economy necessarily assumes that significant activities currently fueled by fossil energy will be converted to electricity: automobiles will run on batteries and homes currently heated with gas or oil will be converted to electric heat pumps. All those EV batteries charging and heat pumps running (largely at places and times that the sun doesn't shine) will add significantly to electricity demand and more than offset any gains in the efficiency of electricity use.

This book focuses on individual consumption and carbon emissions, but it is easy to lose sight of the vast energy consumption and carbon emissions of the industrial economy we take for granted. Industries such as steel and cement production and some long-range transport will need to electrify at the same time we electrify our cars and houses. It may not be out of the question to expect a doubling or tripling of electricity needs in a zero-carbon energy economy.

This is not to say there are no solutions to the climate/energy dilemma—just that there are no complete and perfect solutions. There are bullets in the arsenal to fight climate change. It's just that there are no silver bullets that would allow a zero-carbon economy to be achieved within the twenty years we have with no reduction in energy consumption. We can't rely on some technology that has yet to be invented, because implementation must start now and we don't know for sure that some technological fix will be invented. Despite fifty years of efforts, we still have no cure for cancer—a scientific endeavor that does not have to surmount a fundamental law of physics. The first law of thermodynamics states that energy can neither be created nor destroyed, only stored and transformed. The existing technologies that we have to work with involve environmental, social, and energy compromises. For example, there are other forms of energy storage besides lithium-ion batteries. But existing storage technologies such as pumped storage hydro reservoirs involve higher energy losses than lithium-ion batteries and have severe environmental and social impacts.

Just ask any community that has been ejected from their home to make room for a new hydroelectric power reservoir.

There are a few realistically optimistic studies examining how the United States and other nations might achieve a zero-carbon economy by 2050. Most notably, Stanford researcher Mark Jacobson and colleagues issued a 2017 report entitled "100% Clean and Renewable Wind, Water, and Sunlight All-Sector Energy Roadmaps for 139 Countries of the World." Jacobson et al. map out what it would take to have 100 percent reliance on solar, wind, and hydropower, though this study has been criticized for making unrealistic assumptions about the engineering feasibility of hydropower system modifications. As another example, former New York City mayor and financial news magnate Michael Bloomberg formed a new economic consulting group, New Energy Finance, specifically to assess the prospects for a sustainable energy economy (and to advise enterprises how to profit). In his book, *Climate of Hope* (cowritten with Sierra Club president Carl Pope), he argues for a carbon-free energy grid powered by solar and wind, but with some nuclear power as well. *Climate of Hope* is short on details of how this energy economy would balance renewable supply and industrial and consumer demand. But a 2017 review paper concludes that there are no geophysical constraints on reaching the point where 80 to 90 percent of grid power is supplied by renewable energy. This paper sees difficulty, however, in achieving a 100 percent renewable energy grid; the difference, presumably, would have to be made up by nuclear generation.

What even these cautiously optimistic assessments have in common is at least partial reliance on biomass energy (burning wood pellets for electricity), nuclear power, and/or hydroelectric power to make ends meet, all of which are strongly opposed by many environmental groups. Even wind projects routinely draw the ire of community-based environmental groups. Biomass and hydropower have both been called "worse than coal" for climate by green groups. In short, no one has published a peer-reviewed, no-compromises roadmap for achieving 100 percent renewable replacement power without relying on hydroelectric, biomass, or nuclear energy.

Other thorough assessments are less rosy. David MacKay's book *Sustainable Energy—Without the Hot Air* is a more skeptical assessment of the possibility of implementing 100 percent renewable energy without some basic changes in mobility and energy consumption. A similar assessment of a zero-carbon energy future was released in 2016 by Richard Heinberg and David Fridley of the Post Carbon Institute. This study, *Our Renewable Future*, does not rely heavily on hydroelectric power and rejects nuclear power as a practical solution. The authors take a close look at future energy scenarios, including the problems of embedded energy in solar and wind generation and storage, and conclude that "even assuming a massive buildout of solar and wind capacity during the next 35 years, renewables will probably be unable to fully replace the quantity of energy currently provided by fossil fuels, let alone meet projected energy demand growth." In addition to the economic implications of this change in energy cost and quantity, Heinberg and Fridley say there will be less mobility because of the paucity of renewable liquid fuels for transport. In other words, Americans living in a post-carbon energy world will not be able to drive and fly the distances that they currently take for granted—whether for commutes, work, or pleasure.

HYDROGEN IS A BATTERY, NOT AN ENERGY SOURCE

Not a few people are convinced that hydrogen is the climate-neutral, liquid fuel of the future. Just replace fossil fuels with hydrogen, advocates say, and cars and trucks will have zero-carbon emissions. Toyota is already marketing a hydrogen-fuel-cell-powered car in California, where there are just enough hydrogen fuel stations to make it viable. The key challenge for hydrogen fuel is that pesky first law of thermodynamics: there are no vast reserves of high-energy elemental hydrogen out there waiting to be mined. Currently, hydrogen is produced by stripping natural gas, which releases just as much CO_2 into the atmosphere as burning that gas. Yes, there is hydrogen in water, but this is hydrogen that has already been "burned" (combined with oxygen to make H_2O). To free this hydrogen and make it burnable again, you need a source of energy that is greater than the energy you expect to get back from a fuel cell or combustion. This means that hydrogen fuel is just a way of storing and transporting energy from some other source—renewable or nuclear or fossil. Using electrolysis to produce hydrogen from water is much less efficient than battery technologies like the lithium-ion batteries used in the current crop of EVs. So absent some technological breakthrough, the hydrogen fuel cycle remains just beyond practical limits. And hydrogen fuel still requires a massive buildout of zero-carbon energy sources to power hydrogen production.

Liquid fuels pose a particular challenge in a 100 percent renewable energy regime. Efforts to promote corn-based ethanol fuels have largely resulted in higher food prices and fuels that require substantial amounts of fossil fuel energy to process. Palm oil–based biodiesel demand has led to destruction of tropical mangrove forests, leading to substantial greenhouse gas emissions and loss of the carbon sinks needed to address global warming.

There have been some proposals floated to scrub carbon dioxide from the atmosphere and convert the CO_2 into synthetic hydrocarbon liquid fuels. Synthetic liquid fuels created from atmospheric carbon dioxide are theoretically possible, but only by using substantially more energy than you can get from the resulting liquid fuel. The first law of thermodynamics still applies! The energy would have to be supplied by some acceptable renewable source—wind or solar power. Fuel synthesis might be a practical way to put excess solar and wind power to use when generation exceeds demand. But there will be energy lost in the conversion process: only 60 to 70 percent of the energy input will be stored in the resulting fuel. Soletair, a joint Finnish-German research enterprise, has tested a modular solar-powered, CO_2-stripping liquid fuel plant that fits in a shipping container. This modular unit could turn out about 20 gallons of gasoline or diesel a day. The resulting fuels would have to be expensive to make the capital investment in a part-time facility pay, but they would be carbon neutral. One estimate of the production cost of renewable energy–based synthesized fuels is on the order of $6 per gallon (not considering capital facility investment)—four times more than the $1.50 per gallon current production cost of fossil crude oil.

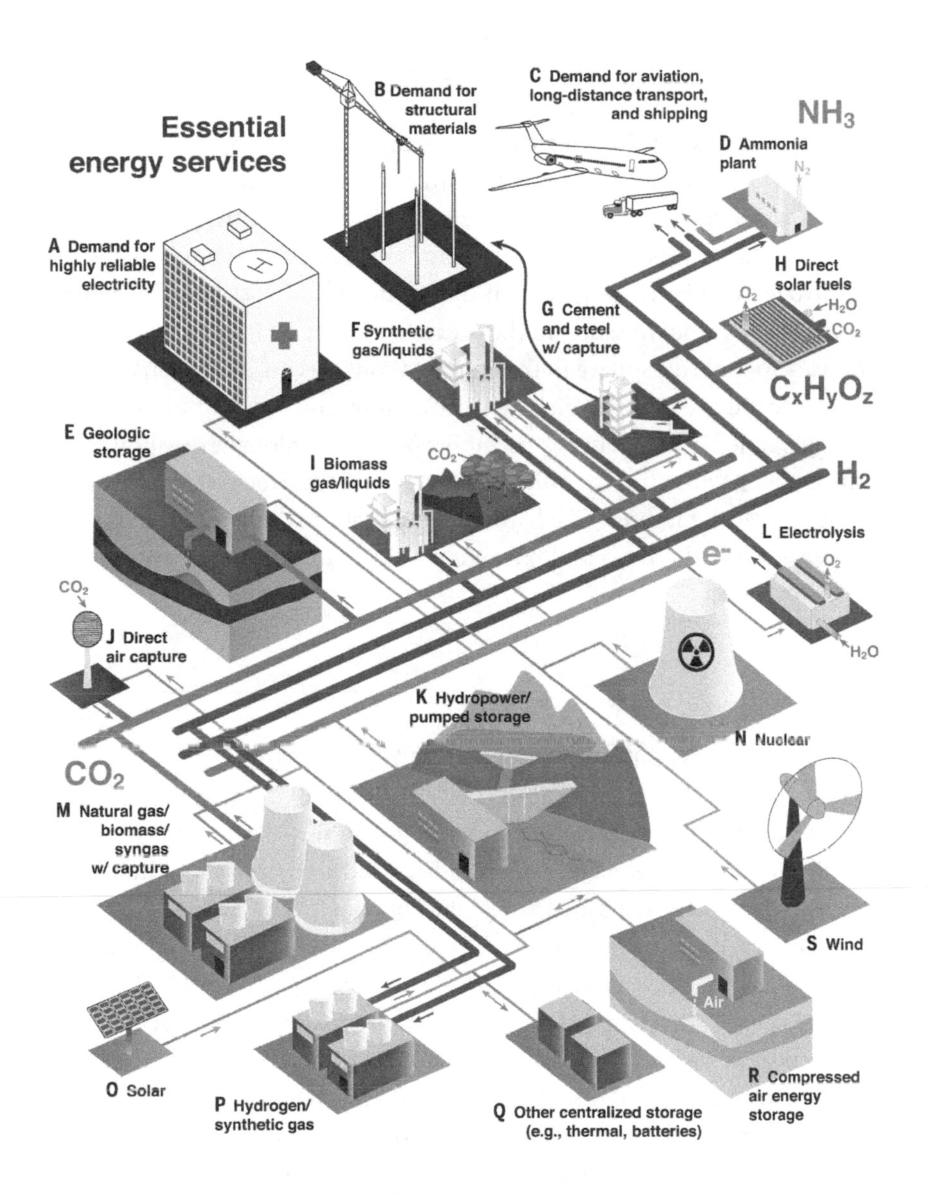

Schematic diagram of zero net carbon emissions energy system.

Source: Steven J. Davis et al., "Net-Zero Emissions Energy Systems," *Science* 360, no. 6396 (June 2018).

Zero-Carbon Food?

Like carbon-neutral liquid fuels, zero-carbon food poses serious challenges. Globally, large-scale agricultural crop production currently depends heavily on synthetic fertilizer produced using the Haber-Bosch method. This process uses natural gas to synthesize ammonia (a molecule of nitrogen and hydrogen) from nitrogen in air and hydrogen in natural gas, using heat and pressure. In the process, the carbon in the natural gas is released to the atmosphere in the form of (you guessed it) carbon dioxide, the principal greenhouse gas. Synthetic ammonia production accounts for 2 percent of global energy use and 1 percent of global greenhouse gas emissions. Although ammonia can also be synthesized using electricity and electrolysis methods like hydrogen generation, this process has yet to be perfected at the rates and efficiencies needed to challenge the Haber-Bosch process.

IS ORGANIC AGRICULTURE ZERO CARBON?

Environmentalists generally support organic agriculture because its rejection of toxic pesticides and synthetic fertilizers promotes human health, long-term soil productivity, and community-based food systems. But can organic agriculture achieve a zero net carbon future? A side-by-side organic versus conventional agriculture comparison project conducted by the Rodale Institute for decades concluded that the organic plot achieved comparable yields to the conventional plot, with about one-third less greenhouse gas emissions. That's an improvement, but it is far from the net zero emissions that agricultural production needs to achieve in the long run. The largest source of greenhouse gases for the conventional plot was the emissions embedded in synthetic fertilizers, while the largest component of greenhouse gas emissions for the organic plot was from the diesel fuel needed to run the farm machinery—more diesel per hectare than was needed for the conventionally farmed acreage. Other studies have found that organic agriculture produces between 5 percent and 34 percent less crops per acre than conventional agricultural methods. Organic fertilizers currently rely heavily on animal manure, so organic agriculture is still tied indirectly to meat and dairy production. But organic fertilizers could be produced from sewage treatment sludges—more palatably referred to as "biosolids"—and legume crops can fertilize soil without animal manure inputs. Fossil diesel fuel might be replaced with renewably sourced synthetic fuels, or electric farm equipment, just as synthetic fertilizers for conventional agriculture might someday be produced using renewable electricity. So organic farming is not by itself a solution for zero-carbon agriculture, but combined with other agriculture and energy mitigation approaches, it can be an important part of the climate solution.

Even larger impacts from agriculture stem from nitrous oxides and carbon released from soil tillage and fertilizer application, as well as from methane from livestock, manure management, and rice cultivation. These agricultural greenhouse gas emissions currently account for 10 percent of total U.S. greenhouse gas emissions. These impacts can be reduced by low-till and no-till agriculture, better fertilizer application management, dry manure management, and periodic draining of rice fields during the growing season. But these emissions cannot be eliminated using better crop management techniques, just reduced. Even if agricultural greenhouse gas emissions are reduced by 50 percent, remaining emissions will pose a substantial obstacle to achieving a zero-carbon agricultural economy.

Only by incorporating carbon sinks into agriculture—essentially setting aside land to sequester the carbon produced by agricultural production—can agriculture hope to achieve zero net greenhouse gas emissions. Achieving zero net food production almost certainly means substantially more acres per unit of crop production, which means that the still-growing global population will have to, on average, eat lower on the food chain. In other words, the diet in a zero-carbon food system will have to be primarily plant based, with relatively little meat—like the Mediterranean diet, as it so happens.

* * *

Forecasting the exact pathway to the needed zero-carbon emissions future is well-nigh impossible, and no one is really in a position to predict the exact social, political, and economic path the future balance of energy production and climate protection will take. But while the details may be unpredictable, at least two points seem inevitable if we act collectively to address greenhouse gas emissions. First, private use of liquid fuels will be rare or even nonexistent, so at the individual level, we will have to electrify everything, including heat and transportation. Second, energy will be less abundant and more precious. It will cost more. So we will have to spend a greater proportion of our incomes on it, use less, or both.

INDIVIDUAL PLANNING FOR A ZERO-CARBON WORLD

What does a zero-carbon economy look like for individual lifestyle choices? We have just twenty years or so to get ready for it. Still, that may seem a long time ahead for individual planning. Most people barely plan a week in advance for individual lifestyle and consumption choices, except for vacation travel, when we might make plans months in advance. Planning for a zero-carbon energy economy that is still decades away may not seem that urgent. But there are some decisions and plans we make as individuals that affect our life for decades. People get married, decide to have children, buy houses with thirty-year mortgages, and (hopefully) plan for children's educational expenses and their own retirement. We routinely buy "durable goods"—particularly cars and household appliances—that we expect to last at least years, if not decades.

Climate-conscious individuals will want to integrate that consciousness into the long-term plans they do make. Individuals committed to the climate sustainability of their own lifestyle should consider whether, taking the pace of technological progress into account, that lifestyle could still be sustained in a zero-carbon energy economy in the two or three decades we have to get there. In many ways, it is easier—and less expensive—to achieve a near-zero-carbon footprint now than it will be in two decades. Right now, a very small percentage of U.S. households make the choice to buy green energy, for example, so there is an ample supply of green energy contracts and renewable energy credits on the market for a relatively low price. But if everyone had to buy renewable energy because it was the only electricity that was sold, then the price would increase substantially until overall energy demand declined to match a more limited supply of 100 percent renewable energy. One study, commissioned on behalf of Citizens Climate Lobby, predicts that a 25 percent increase in electricity prices would be needed to achieve a 50 percent reduction in total U.S. greenhouse gas emissions in twenty years. Achieving zero emissions would necessarily result in a greater price increase.

What does this mean for the "Big Four" parts of our climate footprint—electricity, heat, transport, and food? We can realistically hope that zero net carbon electricity will be available—most likely by combining the variable resources of solar and wind with some measure of dispatchable hydropower, pumped hydro storage, biomass generation, nuclear energy, and improved battery storage. But that electricity will likely be much more expensive than current fossil-generated electricity, as the cost of solar panels and wind turbines that are themselves produced with nonfossil energy goes up. Those living in cold climates will have to trade in their oil- and gas-fired furnaces for some form of electric heat, most likely a ground-sourced heat pump. Higher electricity costs will make thermal efficiency investments more attractive, not because superinsulating your home will become cheaper, but because the higher cost of energy will make the cost savings more attractive. Food, especially meat, is likely to be relatively more expensive, as more acres of land will be needed to support zero-carbon crop production and the cost of fertilizer synthesis with renewable energy will increase.

If you currently drive to work, you probably would still be able to drive, but your car will probably be electric. Higher energy costs might make commuting twice as expensive as it is now; 60-mile one-way commutes might become prohibitively expensive, though one can hope that fully electrified zero-carbon public transit systems will be available.

What does a likely 2040 zero-carbon economy scenario mean for individual decision making in 2020? At a minimum, it should affect long-term choices about where you work and where you live. Buying a house with a thirty-year mortgage in a far-flung suburb with a 60-mile one-way commute and no public transit options is simply a mistake. The average U.S. commute of 8 to 10 miles each way may well be feasible with an electric vehicle or electrified public transit, but a daily 120-mile round trip likely will not. If your employment options in a particular city don't pay well enough to cover house costs in that city, then that city's economy is not sustainable in a zero-carbon economy and you should search out one that is.

But a sense of a zero-carbon future also affects our short-term decision making if we take the concept of sustainability seriously, at least when sustainability is defined to take into account technological progress. The luxuries we choose in life—vacation travel, leisure activities, food beyond subsistence (what this book refers to as "fun" and "feast days")—can only be considered sustainable if we can imagine them as being available to future generations similarly situated to ourselves. If it is impossible to imagine a high-meat diet, frequent air travel, a 100-mile daily commute, and life in a large house heated to 75 in the winter and cooled to 65 in the summer in a zero-carbon future, then it is impossible to justify that lifestyle today as "sustainable" in any sense of the word. On the other hand, if you succeed in reducing your individual direct footprint to 4 tons or so, then you will probably be able to continue that lifestyle in 2040 in a zero-carbon economy.

COOKING WITH GAS IN A ZERO-CARBON WORLD?

As someone who enjoys cooking—and eating—good food, my gas stove is the one appliance I have a hard time giving up for an electric replacement. "Cooking with gas" may be an old advertising slogan, but a gas stove's infinite variability and instant visible response haven't been duplicated by any electric cooktop I am aware of. Plus, there is a great story behind the antique Chambers stove that sits in our kitchen. Robin and I both grew up (Robin in New Hampshire, I in New York) in homes with the exact same model stove, and we never saw another one until we started dating and visited each other's childhood homes. I am hoping that even in a zero-carbon future, there might just be enough renewable biogas to keep the stove. Renewable biogas can be captured from landfills, sewage treatment plants, and agricultural manure processing, and it can also be synthesized (like liquid fuels) with enough renewable electricity. Presumably, it will be much more expensive than fossil gas. But the cooking portion of my current household gas demand is much lower than the portion for space and water heating, so I expect I will be able to afford the luxury of cooking with gas.

I expect that I could continue living in a zero-carbon economy twenty years from now more or less where and how I live now. My daily needs—commuting and heating and lighting my house—are already close to zero net carbon, though these activities will probably become relatively more expensive. I will have to address the remaining big-ticket carbon items in my life: natural gas heat and the fossil fuels we burn for leisure travel to our cabin and on vacation. Electric cars with enough range to get to the cabin are now on the market, so when the time comes to replace our Prius, we will certainly go all electric. When my gas water heater dies, I will have to replace that one with an electric unit as well. At some point, I expect to replace my gas-fired backup heat with a heat pump heating and cooling system, and I expect that as summers get hotter with the level of global warming that is already inevitable, air conditioning will become a necessity. By the time this happens, I expect that heat pump technology—and the practicalities of installing it—will be sufficiently advanced that horror stories like my local friends' stories will be less common. As renewable energy demand increases, I might even have to cut down that ancient walnut tree that shades my roof in order to add rooftop solar to my house. As a consolation, all that walnut wood would heat my house for two or three winters—if a furniture maker doesn't make me a better offer for the tree, and if wood heat is not banned in my area because widespread adoption of wood heat leads to adverse air quality impacts. Predictions are hard to make, especially about the future.

REFERENCES

Bloomberg, Michael, and Carl Pope. *Climate of Hope*. New York: St. Martin's, 2017.

Gerrard, Michael, and John C. Dernbach, eds. *Legal Pathways to Deep Decarbonization in the United States: Summary and Recommendations*. Washington, DC: Environmental Law Institute, 2018.

Heinberg, Richard, and David Fridley. *Our Renewable Future*. Washington, DC: Island Press, 2016.

Jacobson, Mark Z., et al. "Matching Demand with Supply at Low Cost in 139 Countries Among 20 World Regions with 100% Intermittent Wind, Water, and Sunlight (WWS) for All Purposes. *Renewable Energy* 123 (August 2018): 236–248.

MacKay, David J. C. *Sustainable Energy—Without the Hot Air.* Cambridge: UIT Cambridge, 2009.

Risky Business Project. *From Risk to Return: Investing in a Clean Energy Economy.* 2016. http://riskybusiness.org/fromrisktoreturn/.

Shaner, Matthew R., Steven J. Davis, Nathan S. Lewis, and Ken Caldeira. "Geophysical Constraints on the Reliability of Solar and Wind Power in the United States." *Energy & Environmental Science* 11, no. 4 (February 2018): 914–925.

Usher, Bruce. *Renewable Energy.* New York: Columbia University Press, 2019.

CARBON DIARY

Summer 2016

Summertime, and the living is easy! One of the big carbon tabs in life in the Northeast—home heating—goes away. While many people replace their heating tab with a hefty bill for central air conditioning, we never had air conditioning in our house, so we never had to give it up. We keep a nice big fan in the window next to our bed, and few are the nights that are so hot that we can't sleep comfortably with the window fan.

Summer is also the traditional season for outdoor fun and vacation travel. This summer I was assigned to teach Constitutional Law to our second semester law students, so I was not able to arrange a long vacation, but I could take advantage of flexible teaching hours to take some long weekends and a week off here and there.

I started off the month of June in Wilmington, North Carolina, at the annual Waterkeeper Alliance conference. Though this was business travel, I try to take advantage of the carbon invested in my occasional business travel to do some sightseeing and touring as well. There is a group of Waterkeepers who get together for a wilderness float trip wherever the annual conference is, and this year, Hartwell Carson, the French Broad Riverkeeper, hosted our whitewater canoe camping trip in the mountains of western North Carolina. Three nights camping under the stars on the French Broad as it carves and tumbles through the Appalachians made all those travel pounds of CO_2 worth it.

My teaching schedule did not leave time for the canoe trip and an all-day train ride back, so I gritted my teeth and hopped a flight from Charlotte, North Carolina, back to New York, arriving just in time for

The author's sailboat at sea.

Photo courtesy of Karl Coplan.

my class. The carbon impacts of flying that distance—about 260 pounds (based on the carbonfootrpint.com calculator)—turned out to be not that much more than the impacts of taking the train (209 pounds, based on the Union of Concerned Scientists report on Amtrak). We also made a weekend trip to our Adirondack Mountain cabin in June, as well as a sailing trip down the Hudson River to the ocean for some fishing and fresh air. We went canoe camping in the Adirondacks, and, while we caught no fish on the ocean, we had a surprise close encounter with a breaching humpback whale off of Rockaway Point in Brooklyn. My carbon tab for the month was 753 pounds—right on track for a 4-ton annual average, even with the plane and train travel.

In July we enjoyed some more low-carbon weekends on our sailboat—another quick overnight to the ocean over the Fourth of July weekend, and a longer weekend sailing trip to Block Island, Rhode Island.

The Block is a beautiful resort island about 10 miles off the coast of Rhode Island—150 miles by water from our home port in Nyack, New York. Its sparkling waters, clay cliffs, and pristine beaches are a little too far for a weekend trip by sailboat, but we were able to make it out there and back home by Sunday evening by sailing partway the Wednesday before and parking the boat in Stamford, Connecticut, for the night. I could teach my Thursday afternoon class before sailing the rest of the way to the Block, enjoying a marvelous beach weekend, and sailing back on the ocean to the Hudson River in time for my Monday class. My carbon tab for July was just 273 pounds, including our trips to the ocean and the mountains and a weekend drive to Harrisburg, Pennsylvania, to visit our son.

Summer term classes ended in early August. We made one more weekend trip to the mountains. Once classes ended, we sailed our boat down the river one more time and set off for a seven-day coastal passage to Charleston, South Carolina. The plan was to leave the boat in Charleston and return at the end of the winter to take a longer, low-carbon cruise to the Bahamas during my semester break from teaching. Ocean sailing is the true getaway from the carbon-fueled rat race. The wind is free, and even though it blew against us, it blew gently enough that except for the first night, our days and nights of tacking down the coast went by in a summertime blur of clear horizons, dolphins jumping in our wake, and a daily swim break during the midday calms. We arrived safely in Charleston seven restful days after leaving New York Harbor and tied up at the Maritime Center in town. Neither of us had been to Charleston before, so we played tourist on our folding bicycles and sampled the culinary delights at night. After three nights in Charleston, we sailed the boat up the Wando River to the Charleston City Boatyard for haulout. After four August South Carolina nights without air conditioning, we treated ourselves to an overnight at the Inn at Middleton Place and toured its formal gardens. We took the overnight Amtrak back to New York the next night, just in time for fall semester classes. My carbon tab for August was 669 pounds—including the Amtrak travel, a few gallons of diesel we burned getting around Cape Hatteras and into Charleston Harbor, and the Uber rides to and from the train stations.

TAKING STOCK FOR THE YEAR—LIVING WELL WITHIN, AND WELL, WITHIN, MY CARBON BUDGET

September and the arrival of the fall term marked one full year of keeping a daily tally of my carbon footprint and aiming to live on a 4-ton annual direct carbon footprint. So it was time to tally up my carbon emissions for the year and see how I did. I was surprised—and pleased—when I hit the "sum" function on the spreadsheet I was keeping. My direct carbon footprint for the year totaled up to 5,288 pounds, or about 2.7 tons of carbon dioxide equivalents. I had totally underspent my carbon budget for the year, with some margin left over for things I did not try to account for, like dairy and cheese consumption! It turns out I could have added a round-trip cross-country flight somewhere, or (as Robin points out) I could have kept the house warmer last winter.

The biggest source of carbon emissions for the year was driving on our weekend trips in the Prius; my share of this fuel consumption added up to about 1,700 pounds of carbon dioxide, or about the equivalent of one cross-country round-trip plane flight. But that travel included countless trips to the mountains. The next largest item in my carbon budget was natural gas (mostly for heating the house); my share of this added up to about 1,300 pounds of CO_2 for the year. We benefited from a relatively mild winter, so the woodstove and deadwood from the backyard kept the furnace off most of the year. Gas for driving and natural gas for heat took up most of my carbon footprint. Train travel—Amtrak and the occasional commuter rail for trips into New York City—were the other significant items, totaling about 1,000 pounds of CO_2 equivalent.

It was a good year! I don't feel that I had to give up anything important in life to keep within my carbon budget. While I mostly avoided air travel, I made plenty of trips to beautiful beaches, wilderness camping sites, awe-inspiring mountains, and a couple of cities that were new to me, as well as visiting our adult children. I got to work every day with a pretty normal commute most the time (OK, I know most people usually don't paddle a kayak to work). I ate great food and had some new adventures with my spouse and life companion.

POSTSCRIPT

Individual Climate Action in the Trump Era: Now More than Ever

In December 2015, the annual "Conference of the Parties" for the United Nations Framework Convention on Climate Change was held in Paris. Virtually every nation in the world was represented at this meeting. The stakes were high: the 2000 Kyoto Protocol, the first international agreement to set specific national greenhouse gas limits, had expired without being renewed. The U.S. Senate had refused to ratify the Kyoto Protocol when it was negotiated, and the announced opposition of the Senate to any treaty that would bind the United States to greenhouse gas reductions had bleakened the prospects for any international agreement to control global warming. In addition, developed and developing nations were loath to allow any international body to set enforceable limits on energy production and economic activity within their borders. But several events during the previous year set the stage for progress. The Obama administration announced the Clean Power Plan, a series of regulations under the Clean Air Act that would put the United States on a course to reduce power-sector carbon emissions. And the United States, China, India, and Brazil all made pledges to commit to emissions reductions.

On December 12, the Conference of the Parties approved the Paris Agreement. Beaming world leaders at the dais joined raised hands in a victory sign. The assembled nations of the world had agreed to try to limit global warming to 1.5°C and to achieve zero net emissions of greenhouse gases by the middle of this century. Developed and developing nations alike committed to limiting their greenhouse gas emissions. The climate community largely celebrated the agreement, and the world breathed a sigh of relief. Based on the headline news reports, the public might reasonably believe that the world was now on a path to prevent catastrophic global warming. The media seemed to think so; climate coverage all but disappeared during the ensuing year, as coverage of the 2016 presidential campaign season got into full swing.

Not all climate activists celebrated Paris, though. The Paris conference achieved consensus via compromise, and the biggest compromise was the omission of any binding, internationally enforceable limits on any nation's greenhouse gas emissions. Instead, the Paris Agreement allowed each nation to determine its own greenhouse gas reduction goals—called "intended nationally determined contributions," or INDCs. The agreement did not provide for any sanctions against nations that failed to meet their stated goals. And even if every nation met its INDC under the Paris Agreement, the global climate would still be on track to far exceed both the 1.5°C and 2°C warming targets. The combined INDCs simply did not contemplate sufficient reductions to achieve the stated goal of the Paris Agreement.

While the Paris Agreement may have represented only a half-step of progress toward mitigating climate change, it was still a symbolic statement of a global consensus to acknowledge the gravity of the threat posed by anthropogenic global warming. It also provided a framework and hope for a future agreement capable of meeting the climate mitigation goals.

Less than a year later, this global consensus and sense of hope were shattered by the surprise election of Donald Trump as president of the United States. Trump ran on a pledge to abandon the Paris Agreement and resurrect the coal industry by relaxing environmental regulations and eliminating policies favoring renewable energy. Despite losing the popular vote to the candidate pledged to continuing the Obama administration's incremental climate mitigation policies, Trump put together an electoral college majority of conservative, rural, and industrial states. True to his campaign pledges, President Trump announced the United States's withdrawal from the Paris Agreement early in his administration, and his EPA administrator initiated the process of repealing the Clean Power Plan regulations.

Trump's election and the implementation of his policies are demoralizing for individual climate action. Personal action seems largely symbolic anyway, so why make any sacrifice, no matter how small, if the planet is destined to burn in runaway climate change anyway? Adding to the sense of gloom are refocused media attention on climate change and more and more suggestions that it is just too late to do anything.

But even as President Trump announced the U.S. withdrawal from the Paris climate agreement, state and local leaders stepped forward to announce their commitment to carrying out the spirit of the Paris Agreement at more local levels. Led by New York, Miami, Chicago, San Francisco, and Houston, over 350 cities across the country joined a climate alliance that promised to meet the percentage greenhouse gas reductions as pledged by the United States under the Paris Agreement—the so-called "Paris to Pittsburgh" movement. The governors of New York, California, and Washington similarly pledged their states to implement the Paris reduction despite a formal withdrawal by the United States. Twenty-one states have joined the pledge to implement the United States's Paris commitment.

This sort of leadership is exactly the right response at the state and local level to the total abdication of climate leadership at our national level. Climate activists have rightly celebrated this bottom-up approach to climate mitigation in the absence of national leadership. But the ultimate bottom-up action is at the individual level. It remains true that individual carbon reductions cannot, by themselves, prevent dangerous global warming. But it is equally true that pledges by 350 of the tens of thousands of cities across the country cannot, by themselves, prevent dangerous global warming. Just as we applaud climate action at the state and municipal level as a foil to national abdication of the issue, we must support climate action at the most local level of all—in every home in the country.

Individual climate action is now more important than ever. And I don't mean just hopping on a plane to march in a protest or camping out in front of a pipeline carrying fossil fuels destined to fuel the plane you hopped. I mean setting individual climate reduction goals and meeting them. The governors and mayors who are making reduction commitments are not saying, "Gee, why should we deprive ourselves of fossil fuel–powered economies to meet the Paris goals when red states like Alabama and South Carolina are going to keep on fouling the planet?" Instead, they are making the energy future happen now, taking local action even knowing that in the long run, to be effectual, national and international commitments will be required.

The same leadership principles apply at the individual level. You can give up and say, "Why bother reducing my individual footprint when national policy is to bring back coal-fired power plants and increase emissions?" Or you can recognize that social change starts at the individual level and be a climate leader through conspicuous nonconsumption. The 2014 Peoples Climate March was an impressive massing of public sentiment, but not impressive enough to prevail at the polls two years later. Climate activists still suffer from the disconnect between words and actions: we argue that climate change is the gravest threat to humanity since the plague, but we live our lives as if climate change is not worth giving up the creature comforts and mobility that our grandparents were perfectly happy to live without. If effective climate response is to become a political reality, we can't simply make political arguments through protests and marches; we have to convince. And the basic policy argument for an effective climate response is that global warming is a grave—even existential—threat to human civilization as we know it. That argument is not convincing as long as the people making the argument live their lives as if their own contribution to climate change is meaningless.

In January 2019, sixteen-year-old Swedish student Greta Thunberg captivated the world by giving an impassioned address to the World Economic Forum at Davos, Austria. Ms. Thunberg had organized a weekly school strike among the world's youth, a mass protest by those most affected by the loss of hope for the future posed by uncontrolled greenhouse gas emissions. She did not mince words at Davos:

> Our house is on fire. I am here to say, our house is on fire. . . . According to the IPCC . . . we are less than 12 years away from not being able to undo our mistakes. In that time, unprecedented changes in all aspects of society need to have taken place, including a reduction of our CO_2 emissions by at least 50 percent. . . . Solving the climate crisis is the greatest and most complex challenge that *Homo sapiens* have ever faced. The main solution, however, is so simple that even a small child can understand it. We have to stop our emissions of greenhouse gases. . . . Either we do that or we don't.

Thunberg called out the irony of the world's economic leaders flying to the meeting on private jets; she herself arrived by train from Sweden and slept in a tent in the Alpine winter. Thunberg has quit air travel and has refused multiple invitations to give speeches, pointing out that "my generation won't be able to fly other than for emergencies, in a foreseeable future if we are to be the least bit serious about the 1.5° warming limit." She commented to CNN that "you have to practice as you preach; otherwise people won't take you seriously."

Thunberg's school protest has spread to over one hundred countries with thousands of students participating. Climate activism may have found an unlikely communicator in the soft-spoken, shy student—but Thunberg is a leader who is suffering genuine climate loss and who has determined to give up those luxuries in the developed world that contribute to climate change. Like the victims of race discrimination in Montgomery, Alabama, the world's youth are coming to recognize that opting out of a destructive system is more important than personal convenience.

Thunberg started her school strike to force Sweden to live up to its commitments under the Paris Agreement. The Paris Agreement itself is a set of nonbinding, individually determined commitments. For the United States, withdrawal from Paris is completely nonsensical because, if the committed reductions had been too onerous, the United States could have modified its commitments unilaterally, at will. And the U.S. voluntary commitment was never, by itself, enough to limit warming to 2°C. But it was a public declaration of a step in the right direction.

Why don't each of us set our own private Paris goals, like the governors and mayors of the climate alliance? The U.S. "intended nationally determined commitment" (INDC) under the Paris Agreement was a 26 percent reduction by 2025, with a long-term goal of achieving an 80 percent reduction in national emissions by 2050. If states and localities across the country are willing to commit to that reduction, why not each of us personally?

The Paris Agreement asks each nation to take stock, decide for itself how much greenhouse gas reduction is fair and achievable, and implement that reduction. Any individual concerned about the climate can make this same determination. The Paris Agreement contemplated nationally determined carbon reduction goals; individuals can carry out the spirit of the Paris Agreement by committing to "intended *individually* determined contributions" to carbon reduction. Since the 2016 election, I have stuck to my 4-ton annual carbon footprint goal, and I remain committed, as contemplated by the global consensus in Paris, to reach net zero by 2040. Based on what I have learned about low-carbon alternatives, there is no reason I can't reduce that 4-ton budget by 28 percent by 2025; I have already reduced my direct footprint for calendar year 2018 to about 2 tons. We'll drive the family Prius into the ground by 2025, and we can replace it with an electric car with all the range we need to get to our cabin upstate, eliminating the largest piece of my remaining carbon footprint. And my goal is still to achieve zero net carbon by 2040, as we all must.

The Paris reductions were all voluntary. Likewise, we can each make our own voluntary commitment to take part in addressing climate change. These individual actions are as meaningful as actions at the municipal or state level. We can all be part of the Paris Agreement.

APPENDIX

SAMPLE CARBON FOOTPRINT CALCULATION

I used a spreadsheet to keep track of significant greenhouse gas emissions from my daily activities. This is the spreadsheet for August 2016, which includes the annual total for the twelve months from September 2015 through August 2016. Column R contains a formula to calculate and total the CO_2e emissions for each row. I explain the emissions factors I used in notes following the spreadsheet.

A	B	C	D	E	F	G	H	I	J	K	L	M	N	O	P	Q	R
			BUS COMMUTE	DRIVING PRIUS		ELECTRIC VEHICLES	NATURAL GAS		BEEF AND LAMB	AIR TRAVEL	AMTRAK	COMMUTE RAIL	BOAT DIESEL		OTHER		DAILY
DATE	ACTIVITY	NOTES	MILES	MILES	PERSONS	KWH GRID	CCF.	PERSONS		LB. CO_2E	LB. CO_2E	MILES	GAL.	PERSONS	DESCRIPTION	CO_2E	TOTAL CO_2E
8/1/2016	Work at home				1			2						1			0
8/2/2016	Commute to work	Bike and kayak			1			2						1			0
8/3/2016	Commute to work	E-moto			1			2						1			0
8/4/2016	Commute to work, drive to Adirondacks	Smart		215	2			2						1			43
8/5/2016	At cabin				1			2						1	Butane	0.2	0.2
8/6/2016	At cabin				1			2						1	Butane	0.2	0.2
8/7/2016	Drive to trail, North Creek			15.4	2			2						1			3.08
8/8/2016	Drive home from Adiron-dacks			215	2			2						1			43
8/9/2016	Commute to work	Bike and kayak			1			2	0.5					1			13.5
8/10/2016	Commute to work	E-moto			1			2						1			0
8/11/2016	Last-minute cruise errands, setting sail	Smart and Prius		16	2			2					1	2			13.2

8/12/2016	Sailing to Charleston	Under way			1			2						1	Propane for cooking on board	0.1	0.1
8/13/2016	Sailing to Charleston	Under way			1			2						1	Propane for cooking on board	0.1	0.1
8/14/2016	Sailing to Charleston	Under way			1			2						1	Propane for cooking on board	0.1	0.1
8/15/2016	Sailing to Charleston	Under way			1			2						1	Propane for cooking on board	0.1	0.1
8/16/2016	Sailing to Charleston	Under way			1			2						1	Propane for cooking on board	0.1	0.1
8/17/2016	Sailing to Charleston	Under way, motored around Hat-teras			1			2					6	2	Propane for cooking on board	0.1	60.1
8/18/2016	Sailing to Charleston	Under way			1			2						1	Propane for cooking on board	0.1	0.1
8/19/2016	Sailing to Charleston	Under way, motored into harbor			1			2					2	2	Propane for cooking on board	0.1	20.1
8/20/2016	In Charleston	Biking everywhere			1			2						1			0
8/21/2016	Bike to Sul-livans, ferry to Fort Sumter	Bike and ferry			1			2						1	8-mile ferry	20	20

(continued)

A	B	C	D	E	F	G	H	I	J	K	L	M	N	O	P	Q	R
			BUS COMMUTE	DRIVING PRIUS		ELECTRIC VEHICLES	NATURAL GAS		BEEF AND LAMB	AIR TRAVEL	AMTRAK	COMMUTE RAIL	BOAT DIESEL		OTHER		DAILY
DATE	ACTIVITY	NOTES	MILES	MILES	PERSONS	KWH GRID	CCF.	PERSONS		LB. CO_2E	LB. CO_2E	MILES	GAL.	PERSONS	DESCRIPTION	CO_2E	TOTAL CO_2E
8/22/2016	Sail up to Wando, South Carolina	Sail the whole way			1			2					0.5	2	Propane for cooking on board	0.1	5.1
8/23/2016	Uber to Middleton Place	Assumed 1.5 times distance			1			2						1	Uber ride	16.8	16.8
8/24/2016	Uber to train station	Assumed 1.5 times distance			1			2						1	Uber ride	10.2	10.2
8/25/2016	Amtrak to Newark, Uber home				1			2			316.09			1	Uber ride	21.6	337.69
8/26/2016	Commute to work, drive to movie at Lafay-ette Theatre	E-moto! and Smart			1			2						1			0
8/27/2016	Drive kayak to Boat Club, meet Robin at Tallman	Prius		20	2			2						1			4
8/28/2016	Rockland Lake, and fall-term dinner party in Eastchester	Bicycle and Smart			1			2						1			0
8/29/2016	Commute to work	Bike and kayak			1			2						1			0

8/30/2016	Commute to work	E-moto! and Smart			1			2						1			0
8/31/2016	Commute to work	Bike and kayak			1			2						1			0
8/31/2016	August gas				1		13	2						1			78
					1			2						1			0
	Annual totals		316	8432.3	2	55.5	236	2	9.15	260	585.1467	413	27.72701	2		481.4595	5227.686
	lb. CO_2e		42.13333333	1686.46	1		1416	2	256.2				277.2701	1			

For column D, bus commute miles for the Tappan Zee Express bus service were converted into pounds of CO_2e based on an assumed fuel efficiency of 150 passenger miles per gallon (pmpg) and the EPA emissions factor of about 20 pounds of CO_2e per gallon of diesel fuel. The 150 pmpg factor assumes an average ridership of 30 passengers on the bus and a fuel efficiency of 5 miles per gallon (mpg). I compared this calculation with the carbonfootprint.com calculator, and the results were very similar. The formula is CO_2e = (bus miles/150) × 20.

Columns E and F calculate the emissions from rides in our hybrid Prius based on an average fuel efficiency of 50 mpg, divided by the number of people in the car, and then multiplied by the EPA emissions factor of about 20 pounds of CO_2e per gallon of gasoline.

Column G kept track of those times that I charged my electric car or motorcycle away from home, where I have a renewable energy contract for zero-emissions energy. I kept track of how many kilowatt-hours (kWh) I used and multiplied that number by the EPA figure of 0.758 pound of CO_2e per kWh of electricity for the upstate New York electricity grid.

Columns H and I calculated the household emissions from natural gas. I took our monthly gas usage in hundreds of cubic feet (ccf) and multiplied it by 12 pounds of CO_2e per ccf of natural gas; then I divided that number by the number of people living in our house that month, which was usually two. EPA indicates an emissions factor of 0.0551 metric ton of CO_2e per million cubic feet of gas, which translates to about 12 pounds per hundred cubic feet.

Column J kept track of high-impact red meat consumption, using a conversion factor of 27 pounds of CO_2e per pound of these meats. This emissions factor is from the Environmental Working Group study entitled *Meat Eater's Guide to Climate Change and Health* (2011), available at http://static.ewg.org/reports/2011/meateaters/pdf/methodology_ewg_meat_eaters_guide_to_health_and_climate_2011.pdf.

In column K, I kept track of my air travel impacts. For the few times that I flew, I calculated the impacts of the flight using the carbonfooprint.com online calculator and entered the result in pounds of CO_2e.

Column L kept track of Amtrak train travel. Based on the Union of Concerned Scientists study *Getting There Greener: The Guide to Your Lower-Carbon Vacation* (2008), I used an emissions factor of 0.45 pound of CO_2e per mile for the nonelectrified portion of the Amtrak network and 0.37 pound of CO_2e for the electrified portion. I calculated the impacts of each trip by hand and entered the result in pounds of CO_2e.

In column M, I entered the number of miles I rode on commuter rail. This was converted to CO_2e based on a conversion factor of 0.34 pound of CO_2e per mile, taken from the Metropolitan Transportation Authority sustainability report: http://web.mta.info/sustainability/pdf/2012Report.pdf.

In columns N and O, I kept track of the number of gallons of diesel we burned in our sailboat. Gallons of diesel were converted to pounds of CO_2e based on the EPA conversion factor of approximately 20 pounds of CO_2e per gallon, then divided by the number of people on board in column O.

In columns P and Q, I kept track of other carbon-emitting activities that didn't fit into one of the other columns. I calculated the emissions from these activities manually by looking up conversion factors online, or estimating my share of gas or diesel consumption and using the 20 pounds of CO_2e per gallon on the EPA website.

Column R contains a formula that combines the entries and emissions factors for all of the other columns in order to calculate total carbon emissions for the day. The formula is: TOTAL=(D*20/150)+((E*20)/(50*F))+(G*0.758)+ ((H*12)/I)+(J*27)+ K+L+ (M*.34)+(N*20/O)+Q

INDEX

Page numbers in *italics* indicate figures or tables.